THE BEAUTY OF WETLANDS IN HUBEI

大美湖北湿地

湖北省林业局 ◎ 组编

王学雷 张明祥 石道良 ◎ 主编

长江出版传媒 湖北科学技术出版社

图书在版编目（CIP）数据

大美湖北湿地 / 湖北省林业局组编；王学雷，张明祥，石道良主编 . — 武汉：湖北科学技术出版社，2022.10

ISBN 978-7-5706-2232-0

Ⅰ.①大… Ⅱ.①湖… ②王… ③张… ④石… Ⅲ.①沼泽化地—概况—湖北 Ⅳ.① P942.630.78

中国版本图书馆 CIP 数据核字（2022）第 149410 号

大美湖北湿地
DAMEI HUBEI SHIDI

策　　划	章雪峰　邓　涛
责任编辑	宋志阳　傅　玲　万冰怡　邓子林
稿件整理	赵欣梦
装帧设计	胡　博
出版发行	湖北科学技术出版社
地　　址	武汉市雄楚大街 268 号 （湖北出版文化城 B 座 13 ~ 14 层）
邮　　编	430070
电　　话	027-87679468
网　　址	http://www.hbstp.com.cn
印　　刷	湖北新华印务有限公司
开　　本	889mm×1194mm　1/16　20.5 印张
版　　次	2022 年 10 月第 1 版 2022 年 10 月第 1 次印刷
字　　数	300 千字
定　　价	218.00 元

（本书如有印装问题，可找本社市场部更换）

《大美湖北湿地》编委会

主　编

王学雷　　张明祥　　石道良

编　委

杨　超　　吕晓蓉　　康玉辉　　王巧铭　　颜　军

张艳霞　　马梓文　　李中强　　雷　刚　　郝　涛

翻　译　　杨晓晨

摄　　影　（按姓氏笔画排序）

万清平	马正彪	马国飞	王　默	王小平	王英凯
王学雷	王润章	王　瑾	文　林	方纯才	方湘安
卢　文	冯　江	成　英	朱　伟	朱　霁	向三久
刘友华	刘雪芹	孙革会	孙晓东	李万民	李长安
李红海	李　浩	李雪萍	杨　涛	杨敬元	肖华林
肖作华	吴星友	何小白	何秀权	何树林	汪武书
汪常青	张大乐	张心平	张　伟	张传明	张丽亚
张　斌	张　溶	张翼飞	陈凤林	陈　伟	陈庆选
陈忠炎	陈　实	陈　建	邵本刚	武　宏	林　俊
罗晓琳	周学森	郑玉涵	赵广亮	赵欣梦	赵炎柱
赵恩峰	赵　辉	郝　涛	胡良慧	柳　娜	郦希墨
饶　丽	闻　斌	宫少华	袁昌新	莫家勇	贾建国
徐　峥	涂金娥	黄秋生	梅　军	常　飞	崔连芝
崔明启	康玉辉	彭龙生	彭青山	舒仁庆	童长宇
曾江勇	曾晓东	蒲云海	蒙晓东	雷圣祥	雷　刚
雷建高	蔡聪玲	熊　健	颜　军	魏　斌	

前言

湖北省地处长江中游，纵跨长江、江汉两大水系，境内江河纵横交错，湖泊库塘众多，享有"千湖之省"和"鱼米之乡"的美誉，湖北省湿地资源丰富，包含湖泊、河流、沼泽和人工湿地等多种湿地类型。据第二次全国湿地资源调查结果显示，湖北湿地面积达144.5万公顷，占长江经济带湿地总面积的12.5%。目前，湖北省已初步建立起以湿地自然保护区、湿地公园为主体的湿地保护体系，湿地保护率达到52.6%。

习近平生态文明思想是我国改革开放以来生态文明建设的最新成果，也是推进我国生态文明建设迈上新台阶的根本遵循和行动指南。深刻理解和领悟习近平生态文明思想指导下的湿地保护工作，对全面推进我国湿地保护的高质量发展具有重要的意义。我国于1992年7月31日加入《湿地公约》并成为第67个缔约方。履约30年以来，在党和国家的高度重视下，在林草部门的牵头组织下，在各部门和全社会大力支持及国际合作参与下，我国湿地保护取得显著成效。

湖北省始终以习近平总书记视察长江经济带重要指示为根本遵循，坚持以习近平总书记重要论述"山水林田湖草沙是一个生命共同体"为指导思想，湖北作为长江经济带发展国家战略的重要支点和重要省份之一，承担着长江经济带高质量发展的光荣责任。推进长江大保护、建设美丽长江，既是关系国家发展全局的重大战略，也是确保湖北可持续发展的必然选择。湖北省践行"绿水青山就是金山银山"的理念，坚定不移地走"生态优先、绿色发展"之路，始终把修复长江生态环境摆在首要位置，共抓大保护，不搞大开发。将绿色崛起作为湖北高质量发展的重要底色，谱写长江经济带高质量发展新篇章，湿地保护成绩斐然，为长江大保护交出合格答卷。

《大美湖北湿地》一书阐述了我国生态文明建设与湿地保护的关系，总结了我国履行《湿地公约》30年以来湿地保护成就；从湖北湿地资源演变、湿地功能、湿地生态修复、科研宣教、湿地保护体系和制度建设等多个方面，展示了长江大保护背景下湖北湿地保护的行动和实践举措，以及湖北省生态文明建设中湿地保护成果；提出了在新形势下湖北湿地保护和利用的战略展望；同时回顾历史，总结经验，开拓创新，展望未来，展示了湖北湿地保护管理创新探索的典型案例和示范经验，以及在湿地保护领域具有影响的科学家先进事迹，湿地管理者、志愿者的工作贡献；还介绍了参与湖北湿地保护的公益性社会组织以及中小学湿地保护教育等特色内容。在展示湖北湿地资源特点和保护管理成就的同时，也可为全国湿地保护管理工作提供借鉴。

编者

2022年8月

PREFACE

Hubei Province is located in the middle reaches of the Yangtze River, straddling the two major water systems—the Yangtze River and the Hanjiang River. The rivers in the territory are crisscrossed, and there are many lakes and reservoirs, which allows the province to enjoy the reputation of the "province of a thousand lakes" and the "hometown of fish and rice". The wetland resources abound in the region, including lakes, rivers, marshes, artificial wetlands and other types of wetlands. According to the results of the second national wetland resources survey, the wetlands in Hubei occupies 1.445 million hm^2, accounting for 12.5% of the total wetland area of the Yangtze River Economic Belt. At present, Hubei Province has initially established a wetland conservation system with wetland nature reserves and wetland parks as the main body, reaching 52.6% of the wetland conservation rate.

Xi Jinping's thought of ecological civilization is the latest achievement in the construction of ecological civilization since China's reform and opening up, and also the fundamental guidelines and action guides for advancing the construction of ecological civilization in China to a new level. Deep understanding and comprehension of wetland conservation under the guidance of Xi Jinping's thought of ecological civilization is of great significance in comprehensively promoting the high-quality development of wetland conservation in China. China joined the *Ramsar Convention* on July 31, 1992 and became the 67th party. Since 30 years of compliance, under the high attention of the Party and the State, the leading organization of the forestry and grass sector, the strong support of other sectors and the whole society and the participation of international cooperation forces, China's wetland conservation has achieved remarkable results.

Hubei Province has always taken the important instructions from General Secretary Xi Jinping's inspection of the Yangtze River Economic Belt as the fundamental guidelines, and adhered to the important exposition of General Secretary Xi Jinping that "mountains, rivers, forests, farmlands, lakes, grasslands and sands are a life community" as the guiding ideology. As an important pivot point and one of the key provinces in the national strategy for the development of the Yangtze River Economic Belt, Hubei assumes the glorious responsibility for the quality development of the Yangtze River Economic Belt. Promoting the protection of Yangtze River and building a beautiful Yangtze River, is not only a major strategy for the overall development of the country but also an inevitable choice to ensure the sustainable development of Hubei. Hubei Province practices the concept of "lucid waters and lush mountains are invaluable assets" and unswervingly takes the path of "prioritizing ecological conservation and boosting green development", always considering the restoration of the ecological environment of the Yangtze River as a major priority, promoting environmental conservation and avoiding excessive development. The green rise as an important color of the high-quality development of Hubei, writes a new chapter of high-quality development of the Yangtze River Economic Belt. The achievements of wetland conservation will deliver a qualified answer sheet for the the well-coordinated conservation of the Yangtze River.

The book *The Beauty of Wetlands in Hubei* elaborates the relationship between the construction of ecological civilization and wetland conservation in China, summarizes the achievements of wetland conservation since China has fulfilled the *Ramsar Convention* for 30 years; from the evolution of wetland resources, wetland functions, ecological restoration of wetlands, scientific research, publicity and education, the improvement of wetland conservation systems and institutions, it shows the actions and practical initiatives of wetland conservation in Hubei under the background of well-coordinated conservation of the Yangtze River, and the achievements of wetland conservation during the construction of ecological civilization in Hubei Province. This book also puts forward the strategic outlook of wetland conservation and utilization in Hubei in the new situation and context. At the same time, by reviewing the history, summarizing the experience, blazing new trails and looking ahead, this book unfolds the typical cases and demonstration experiences of wetland conservation and management innovation exploration in Hubei, as well as the advanced deeds of influential scientists in wetland conservation, the work contribution of wetland administrators and volunteers, and the public welfare social organizations involved in wetland conservation in Hubei, wetland primary and secondary schools and other special contents. While showcasing the characteristics of wetland resources and conservation management achievements in Hubei, it can also provide a reference for the national wetland conservation and management .

Authors
August 2022

目 录

第一篇　生态文明与湿地保护　/ 001

一、人类生态文明的载体——湿地　/ 004

二、我国生态文明建设战略与湿地保护　/ 007
1. 我国的生态文明建设战略　/ 007
2. 我国的湿地保护战略　/ 007

三、习近平总书记高度重视湿地保护　/ 013
1. 习近平生态文明思想指导下的湿地保护　/ 013
2. 习近平总书记针对湿地保护的重要讲话　/ 019

四、我国湿地保护成效　/ 022
1. 持续推进湿地保护修复工作　/ 022
2. 不断健全湿地保护法规和制度体系　/ 022
3. 不断完善湿地保护机构和体系建设　/ 026
4. 稳步推动湿地调查监测体系建设　/ 026
5. 不断深化湿地履约和国际合作　/ 027

第二篇　长江大保护背景下湖北湿地保护实践　/ 043

一、湖北省湿地环境演变与资源　/ 044
1. 湖北湿地形成与演变　/ 044
2. 湖北水系与河湖湿地　/ 047
3. 湖北湿地资源概况　/ 048
4. 湖北湿地文化资源　/ 053

  5. 独特的湿地景观 / 055

二、"千湖之省"湖北湿地功能 / 056

  1. 湖北湿地资源功能 / 056

  2. 湖北湿地生态功能 / 058

  3. 湖北湿地人文功能 / 061

三、湖北湿地保护行动与措施 / 061

  1. 践行习近平生态文明思想，以新理念引领湿地保护发展 / 061

  2. 完善湿地保护顶层设计，用最严格最严密的法治保护湿地 / 062

  3. 创新机制体制，强化湿地保护管理的责任与绩效考核 / 063

  4. 全面实行河湖长制、林长制，推进河湖湿地治理体系和
治理能力现代化 / 067

  5. 强化科技支撑，大力开展湿地保护修复 / 069

  6. 积极推进湿地保护国际合作 / 070

四、湖北湿地保护管理的重要成效 / 070

  1. 湿地生态修复成效明显 / 070

  2. 湿地保护管理评价体系基本建立 / 072

  3. 开展合理利用湿地特色资源 / 073

  4. 社会公众的湿地保护意识增强 / 078

  5. 成功申办《湿地公约》第十四届缔约方大会 / 079

第三篇　湖北湿地保护利用战略展望 / 103

一、以《湿地保护法》实施为契机，全方位推动湖北湿地保护管理 / 105

  1. 加强湿地普法宣传，提高法治意识 / 105

  2. 构建分级体系，完善制度和机制建设 / 105

  3. 科学规划，全面保护修复湿地 / 105

  4. 实践创新，打造新型"湿地保护+"模式 / 105

  5. 立足发展，实现湿地保护与产业经济协同共生 / 106

  6. 加强监督管理，落实河湖长制和林长制的湿地保护责任 / 106

  7. 更新理念，持续发力，湖北湿地保护上新台阶 / 110

二、强化法治化建设，加快颁布《湖北省湿地保护条例》 / 110

三、完善智慧湿地系统，综合管理能力上新台阶 / 111

四、持续推进国际合作，促进国际重要湿地和湿地城市建设 / 112

第四篇　湖北湿地保护管理典型案例 / 121

一、武汉国际湿地城市——城市湿地保护与综合管理 / 124

  1. "百湖之市"的武汉 / 124

  2. 武汉湿地保护地体系建设 / 125

  3. 武汉湿地保护管理制度和创新机制 / 127

  4. 实施重要湿地保护和恢复工程 / 129

二、天鹅洲长江故道湿地——珍稀物种保护地管理探索 / 132

  1. 长江中游故道群湿地的特殊性和典型性 / 132

  2. 国家级自然保护区的建设和管理 / 133

  3. 故道群湿地与濒危物种迁地保护 / 137

三、神农架国家公园——湿地保护与管理创新探索 / 138

  1. 神农架国家公园生态定位与区域湿地特色 / 138

  2. 神农架生态建设与湿地保护 / 140

  3. 社区共建创新机制 / 140

四、洪湖国际重要湿地——湖泊保护与综合治理实践 / 145

  1. 洪湖湿地生态区位与功能 / 145

  2. 洪湖湿地保护恢复与综合管理 / 146

五、丹江口水库——重要水源地与流域生态保护 / 150

  1. 丹江口库区湿地及水资源 / 150

  2. 丹江口库区水源地保护与生态补偿 / 151

六、襄阳汉江湿地保护联动，流域保护协同机制创先　　/ 153
　　1. 创新汉江保护协作机制是汉江湿地保护之要　　/ 156
　　2. 保持汉江水系连通性和流动性是汉江湿地健康之源　　/ 157
　　3. 开展十大战役等环境整治行动是汉江湿地安全之基　　/ 157
　　4. 恢复和改善汉江原生态是汉江湿地发展之策　　/ 157
　　5. 打造人民群众共享的绿色空间是汉江湿地惠民之路　　/ 157

七、沉湖国际重要湿地——智慧湿地提升科学管理水平　　/ 157
　　1. 沉湖国际重要湿地及自然保护区　　/ 157
　　2. 沉湖生物多样性监控与智慧湿地建设　　/ 158

八、精准布局创新亮点，三位一体聚合效益——天门张家湖湿地　　/ 160
　　1. 创新管理体系，构建全面保护支撑力　　/ 161
　　2. 创本土特色科普，引公众意识入胜　　/ 163
　　3. 创湿地经济品牌，建生态产业基地　　/ 166

九、传承发扬生态圩田智慧，再现孝感朱湖梦里水乡　　/ 167
　　1. 朱湖特色的退田还湖与生态修复之路　　/ 167
　　2. 朱湖湿地圩田与湖乡生态产业　　/ 168

十、远安沮河——全域湿地保护与乡村振兴融合管理　　/ 170
　　1. 远安沮河湿地与资源特色　　/ 171
　　2. 全域湿地保护与管理创新　　/ 172
　　3. 推动小微湿地保护，融合乡村振兴和全域旅游　　/ 173

十一、践行两山论——探索徐家河三生融合高质量绿色发展之路　　/ 175
　　1. 徐家河湿地功能特色与水源地保护　　/ 175
　　2. 三生融合发展理念下社区共建模式　　/ 176
　　3. 高质量迈向湿地全面保护和乡村振兴　　/ 178

十二、汉川汈汊湖——退垸还湖重现自然生态美景　　/ 179
　　1. 汈汊湖演变与湿地现状　　/ 179

2. 汈汊湖退垸（田、渔）还湖与生态修复　　/ 180

　　3. 汈汊湖生态修复与治理成效　　/ 181

十三、绿色赤壁——湿地保护与区域发展新实践　　/ 182

　　1. 赤壁市域典型湖泊湿地　　/ 182

　　2. 赤壁湿地保护管理发展实践　　/ 183

　　3. 赤壁湿地生态保护成效　　/ 185

十四、城市湖泊生态修复，打造湖北湖泊治理样板　　/ 186

　　1. 城市湖泊生态修复的"内沙湖模式"　　/ 187

　　2. 城市湖泊生态修复的"东湖模式"　　/ 191

　　3. 城市湖泊生态修复的"沙湖模式"　　/ 191

第五篇　湖北湿地保护典型人物与组织　　/ 233

一、毕力一生，守护"地球之肾"——著名湿地科学家蔡述明　　/ 234

二、致力于长江湿地保护的"山水教授"——"荆楚楷模"李长安　　/ 239

三、中国第一家——湖北省湿地保护基金会　　/ 242

　　1. 组织开展与湿地相关的普法宣讲、生态公益活动　　/ 243

　　2. 开展湿地宣教、培训活动，提高保护意识与管理能力　　/ 245

　　3. 开展湿地保护修复、资源监测及专家技术支撑　　/ 245

　　4. 加强自身建设和社会参与度，提升基金会的影响力　　/ 247

　　5. 组织社会公益募捐，管理湿地保护基金　　/ 247

四、致力于生物多样性保护的NGO——世界自然基金会　　/ 247

　　1. WWF与生物多样性保护　　/ 247

　　2. WWF与湖北湿地保护　　/ 249

五、湿地保护其他社会组织和人物　　/ 249

　　1. 湖北省野生动植物保护协会　　/ 249

　　2. 武汉观鸟会——志愿者之家　　/ 250

3. 湿地学校——生态教育进校园　　/ 252
　　4. 湿地保护一线的管理人员、志愿者　　/ 260

六、湖北湿地保护的科研支撑　　/ 263
　　1. 中国科学院精密测量科学与技术创新研究院　　/ 263
　　2. 中国科学院水生生物研究所　　/ 265
　　3. 武汉大学梁子湖湖泊生态系统国家野外科学观测研究站　　/ 265
　　4. 中国科学院武汉植物园和中国地质大学（武汉）湿地演化与
　　　生态恢复湖北省重点实验室　　/ 265
　　5. 湖北省林业科学研究院（中国林业科学研究院湖北分院）　　/ 266
　　6. 湖北省野生动植物保护总站　　/ 266

附录：湖北省湿地名录

　　国际重要湿地名录　　/ 297
　　国家重要湿地名录　　/ 297
　　省级重要湿地名录　　/ 297
　　国家级自然保护区名录　　/ 299
　　省级自然保护区名录　　/ 300
　　国家湿地公园名录　　/ 300
　　省级湿地公园名录　　/ 302

CONTENTS

Chapter 1 Ecological Civilization and Wetland Conservation / 028

1.1 Wetlands, the vehicle for human ecological civilization / 028

1.2 China's construction strategy of ecological civilization and wetland conservation / 030

1.2.1 China's construction strategy of ecological civilization / 030

1.2.2 China's wetland conservation strategy / 031

1.3 General Secretary Xi Jinping attaches great importance to wetland conservation / 033

1.3.1 Wetland conservation under the guidance of Xi Jinping's thought of ecological civilization / 033

1.3.2 General Secretary Xi Jinping's important speeches on wetland conservation / 037

1.4 Progress of wetland conservation in China / 040

1.4.1 Continuously improve wetland conservation and restoration / 040

1.4.2 Continuously perfect wetland conservation regulations and institutioanl systems / 040

1.4.3 Continuously promote construction of wetland conservation institutions and systems / 041

1.4.4 Steadily advance the construction of wetland survey and monitoring system / 041

1.4.5 Continuously deepen wetland compliance and international cooperation / 042

Chapter 2 Hubei Wetland Conservation Practices in the Context of Yangtze River Coordinated Conservation / 080

2.1 Environmental evolution and resources of wetlands in Hubei Province / 080

2.1.1 Formation and evolution of wetland in Hubei / 080

2.1.2 Water system and riverine wetlands in Hubei / 081

2.1.3 Overview of wetland resources in Hubei / 082

2.1.4 Cultural resources of wetlands in Hubei / 083

2.1.5 Unique wetland landscape / 085

2.2 Functions of wetlands in Hubei—"province of a thousand lakes" / 086

2.2.1 Resource function of wetlands in Hubei / 086

2.2.2 Ecological function of wetlands in Hubei / 087

2.2.3 Humanistic function of wetlands in Hubei / 088

2.3 Conservation action and measures of wetlands in Hubei / 089

2.3.1 Practice Xi Jinping's thought of ecological civilization and lead the development of wetland conservation with new concepts / 089

2.3.2 Improve the top-level design of wetland conservation and protect wetlands with the strictest and most stringent rule of law / 090

2.3.3 Innovate mechanism and system and increase accountability and performance assessment of wetland conservation and management / 092

2.3.4 Fully implement the river and lake chief scheme and promote the modernization of the riverine wetlands governance system and governance capacity / 093

2.3.5 Strengthen scientific and technological support and vigorously carry out wetland conservation and restoration / 095

2.3.6 Actively promote international cooperation in wetland conservation / 097

2.4 Important results of wetland conservation and management in Hubei / 097

2.4.1 Clear results of wetland ecological restoration / 098

2.4.2 Basic establishment of evaluation system of wetland conservation management / 098

2.4.3 Deepening rational use of wetland resources / 100

2.4.4 Heightened public awareness of wetland conservation / 101

2.4.5 Successful bid to host the 14th Conference of the Contracting Parties to the Ramsar Convention / 102

Chapter 3 Outlook of Wetland Conservation and Utilization Strategy in Hubei / 114

3.1 Take the implementation of the Wetland Conservation Law as an opportunity to promote the conservation and management of wetlands in Hubei in all aspects / 115

3.1.1 Increase publicity of wetland laws, raise awareness of the rule of law / 115

3.1.2 Build a grading system, improve the construction of system and mechanism / 115

3.1.3 Plan in a scientific manner, comprehensively conserve and restore wetlands / 115

3.1.4 Practice innovation, create a new model of "wetland conservation +" / 116

3.1.5 Focus on development, achieve synergy between wetland conservation and industrial economy / 116

3.1.6 Strengthen supervision and management, bear responsibilities of wetland conservation of river and lake chief scheme and forest chief scheme / 116

3.1.7 Update concept with sustained efforts, lift wetland conservation in Hubei up to a new level / 117

3.2 Strengthen legal system and speed up the promulgation of Regulations on Wetland Conservation in Hubei Province / 117

3.3 Improve intelligent wetland system to lift comprehensively management capacity up to a new level / 118

3.4 Promote continously international cooperation to advance construction of wetlands of international importance and wetland cities / 119

Chapter 4 Typical Cases of Wetland Conservation and Management in Hubei / 192

4.1 Urban wetland conservation and integrated management in Wuhan, an international wetland city / 192

4.1.1 Wuhan, the "city of a hundred lakes" / 192

4.1.2 Construction of wetland reserve systems in Wuhan / 193

4.1.3 Wetland conservation and management systems and innovative mechanisms in Wuhan / 194

4.1.4 Implemented projects of important wetland conservation and restoration / 195

4.2 Management exploration on protected areas of rare species in wetlands of Tian-e-zhou Oxbow of the Yangtze River / 196

4.2.1 Special characteristics and typicality of the wetlands in oxbows in the middle reaches of the Yangtze River / 196

4.2.2 Construction and management of national nature reserves / 196

4.2.3 Conservation of wetlands in oxbows and ex-situ conservation of endangered species / 198

**4.3 Innovation exploration on wetland conservation and
     management in Shennongjia National Park** / 200

4.3.1 Ecological positioning and regional wetland characteristics of
      Shennongjia National Park / 200

4.3.2 Ecological construction and wetland conservation in Shennongjia / 201

4.3.3 Innovative mechanisms of community co-construction
      in ecological conservation / 202

**4.4 Practices of lake conservation and integrated governance in wetlands of
     international importance in Honghu Lake** / 203

4.4.1 Ecological location and functions of wetlands in Honghu Lake / 203

4.4.2 Conservation, restoration and integrated management of
      wetlands in Honghu Lake / 203

**4.5 Danjiangkou reservoir, and important water source and conservation of
     its basin ecology** / 205

4.5.1 Wetlands and water resources in Danjiangkou reservoir area / 205

4.5.2 Water source conservation and ecological compensation
      in Danjiangkou reservoir area / 206

**4.6 Integrated efforts for wetland conservation of the Hanjiang River in Xiangyang,
     collaboration mechanism of basin conservation as a pioneer** / 207

4.6.1 An innovative collaborative mechanism for the conservation of the Hanjiang River,
      the key to wetland conservation of the Hanjiang River / 208

4.6.2 Maintaining connectivity and mobility of the Hanjiang River water system,
      the source of the health of the wetlands of the Hanjiang River / 209

4.6.3 Taking environment improvement actions such as ten major campaigns,
      the basis of wetland security of the Hanjiang River / 209

4.6.4 Restoring and improving the original ecology of the Hanjiang River,
      the development strategy for wetlands of the Hanjiang River / 210

4.6.5 Creating a green space shared by the people, the way to benefit
      the people of the wetlands of the Hanjiang River / 210

**4.7 Wetlands of international importance of Chenhu Lake, intelligent wetlands
     with a lifted level of scientific management** / 210

4.7.1 Wetlands of international importance and the nature reserve in Chenhu Lake / 210

  4.7.2　Biodiversity monitoring and construction of intelligent wetlands of
     Chenhu Lake　/ 211
**4.8　Precise layout of innovative beauty and three-in-one aggregation
   benefits- wetlands in Zhangjia Lake of Tianmen**　/ 212
  4.8.1　Innovate management system, build a support force of
     comprehensive conservation　/ 212
  4.8.2　Create local characteristics of science popularization, attract public awareness　/ 213
  4.8.3　Create a brand of wetland economy, build a base of ecological industry　/ 214
**4.9　Carry forward the wisdom of ecological polder to
   recreate dream water town of Zhuhu in Xiaogan**　/ 214
  4.9.1　Path with Zhuhu characteristics of returning land for farming to lakes and
     ecological restoration　/ 215
  4.9.2　Wetland polder of Zhuhu Lake and eco-industry of lake township　/ 216
**4.10　Integrated management of all-for-one wetland conservation and
    rural revitalization area of Juhe River in Yuan'an**　/ 218
  4.10.1　Wetlands of Juhe River in Yuan'an and characteristics of resources　/ 218
  4.10.2　Wetland conservation and management innovation for the whole area　/ 218
  4.10.3　Promoting conservation of small and micro wetlands to
      integrate rural revitalization and all-for-one tourism　/ 219
**4.11　Practicing the theory of two mountains, exploring the path of quality green
    development of three-life integration in Xujiahe River**　/ 220
  4.11.1　Characteristics of wetland functions in Xujiahe River and
      water source conservation　/ 220
  4.11.2　Model of community co-construction under the concept of integrated
      development of three lives　/ 222
  4.11.3　Towards comprehensive wetland conservation and rural revitalization
      with high quality　/ 223
**4.12　Returning land for levees to lakes to restore the natural ecological beauty of
    Diaocha Lake in Hanchuan**　/ 224
  4.12.1　Evolution of Diaocha Lake and its current situation of wetlands　/ 224
  4.12.2　Returning land for levees (farming and fishing) to lakes and ecological restoration of

Diaocha Lake	/ 225
4.12.3 Ecological restoration and treatment effectiveness of Diaocha Lake	/ 226

4.13 New practices of wetland conservation and regional development in green Chibi / 227

4.13.1 Typical lake wetlands in Chibi city area	/ 227
4.13.2 Practices of wetland conservation and management development in Chibi	/ 228
4.13.3 Effectiveness of ecological conservation of wetlands in Chibi	/ 229

4.14 Ecological restoration of urban lakes to create a treatment model of lakes in Hubei / 229

4.14.1 "Neisha Lake model" of ecological restoration of urban lakes	/ 230
4.14.2 "East Lake model" of ecological restoration of urban lakes	/ 231
4.14.3 "Shahu Lake model" for ecological restoration of urban lakes	/ 231

Chapter 5 Typical Figures and Organizations of Wetland Conservation in Hubei / 268

5.1 Cai Shuming, a famous wetland scientist exerting lifelong effort to guard the "kidneys of the earth" / 268

5.2 Li Chang'an, a "Jingchu model" and "professor of mountains and rivers" committing to the conservation of wetlands of the Yangtze River / 272

5.3 Hubei Wetland Conservation Foundation, China's first of its kind / 276

5.3.1 Organize and carry out wetland-related popularization of law and activities of ecology and public welfare	/ 278
5.3.2 Carry out wetland popularization and education and training activities to improve conservation awareness and management capacity	/ 279
5.3.3 Carry out wetland conservation and restoration, resource monitoring and expert technical support	/ 280
5.3.4 Strengthen its own construction and social participation to enhance the influence of the Foundation	/ 281
5.3.5 Organize social welfare fundraising and manage wetland conservation fund	/ 282

5.4 WWF, a NGO dedicated to biodiversity conservation / 282

5.4.1 WWF and biodiversity conservation	/ 282

5.4.2 WWF and wetland conservation in Hubei / 282

5.5 Other social organizations and figures in wetland conservation / 283

5.5.1 Hubei Wildlife Conservation Association / 283

5.5.2 Wuhan Bird Watching Society, home of volunteers / 284

5.5.3 Wetland schools allow ecological education in schools / 285

5.5.4 Managers and volunteers at the front line of wetland conservation / 291

5.6 Scientific research support for wetland conservation in Hubei / 292

5.6.1 Innovation Academy for Precision Measurement Science and Technology, Chinese Academy of Sciences / 293

5.6.2 Institute of Hydrobiology, Chinese Academy of Sciences / 294

5.6.3 Wuhan University and the national field scientific observatory and research station of lake ecosystem of Liangzi Lake / 294

5.6.4 Wuhan Botanical Garden, Chinese Academy of Sciences, China University of Geosciences (Wuhan) and Hubei Provincial Key Laboratory of Wetland Evolution and Eco-Restoration / 295

5.6.5 Hubei Academy of Forestry (Hubei Branch of Chinese Academy of Forestry) / 295

5.6.6 Hubei Provincial General Station for Wildlife Conservation / 296

Appendix: List of Wetlands in Hubei Province / 297

List of International Important Wetlands / 297

List of National Important Wetlands / 297

List of Provincial Important Wetlands / 297

List of National Nature Reserves / 299

List of Provincial Nature Reserves / 300

List of National Wetland Parks / 300

List of Provincial Wetland Parks / 302

第一篇

生态文明与
湿地保护

CHAPTER 1

ECOLOGICAL CIVILIZATION
AND WETLAND
CONSERVATION

清江　Qingjiang River

湿地是地球生态系统的重要组成部分，与森林、海洋共同构成了地球三大生态系统，有着不可替代的系统功能，被誉为"地球之肾"。湿地可为人类提供淡水资源、食物来源，净化水环境、维持生态平衡、减缓气候变化等，与人类的生存、繁衍、发展息息相关，是人类实现生态文明的重要载体。党的十八大把生态文明建设纳入中国特色社会主义事业"五位一体"总体布局，湿地保护是生态文明建设的重要内容，事关国家生态安全，事关经济社会可持

续发展，事关中华民族子孙后代的生存福祉。我国湿地保护先后经历了摸清家底和夯实基础（1992—2003年）、抢救性保护（2004—2015年）、全面保护（2016—2021年）三个阶段。2022年6月1日起，《中华人民共和国湿地保护法》（以下简称《湿地保护法》）正式施行。这是为贯彻落实党中央决策部署、践行习近平生态文明思想，进一步强化湿地生态系统保护与恢复而制定出台的重要立法成果，同时标志着我国的湿地保护迈入高质量发展的新阶段。

长江三峡大坝　　Yangtze River Three Gorges Dam

一、人类生态文明的载体——湿地

湿地是地球上生产力最高的生态系统之一。据估计，全球湿地生态系统每年提供的生态系统服务价值高达 47 万亿美元。为保护全球湿地及湿地资源，1971 年 2 月 2 日，来自 18 个国家的代表在伊朗海滨小镇拉姆萨尔共同签署了《关于特别是作为水禽栖息地的国际重要湿地公约》(简称《湿地公约》，又称《拉姆萨尔公约》)，旨在通过各缔约方的实际行动与国际合作，共同保护和合理利用湿地，为全球可持续发展做出贡献。

《湿地公约》中将湿地定义为"天然或人工的、永久性或暂时性的沼泽地、泥炭地或水域，蓄有静止或流动的淡水、微咸或咸水水体，包括低潮时水深不超过 6 米的海域。"根据《湿地公约》的分类体系，世界范围的湿地可以大致划分为天然湿地和人工湿地。天然湿地又可以进一步划分为近海与海岸湿地和内陆湿地。海岸带区域的珊瑚礁、海草床、红树林、滩涂、盐沼等都是近海与海岸湿地的代表；

而湖泊、河流、淡水草本沼泽、泥炭地等则是典型的内陆湿地。常见的人工湿地有水塘、盐田、水库、稻田等。

纵观古今，湿地与人类文明发展息息相关。闻名人类史的古埃及、古巴比伦、古印度和中国古代文明都是在湿地中孕育并繁衍壮大的。世界上许多河流湿地都是孕育人类原始文明的"摇篮"。在人类生产力水平低下的农耕文明时期，人们通常选择气候适宜、水源充沛、土地肥沃的地区耕作、生活并建立聚居区。在古代中国，黄河和长江滋养了华夏子孙，我们的祖先依傍这两条河流创造了古老而灿烂的中华文明。黄河和长江充沛的水资源和两岸肥沃的土地为农牧业的发展提供了前提条件，中华民族的文明史，就这样伴随着湿地而发源，伴随着汩汩的水声缓缓滋长，润泽而绵长。

在经历了原始文明和农耕文明的发展后，伴随着工业革命，人类迎来了工业文明，在这一时期，人类对大自然展开了空前规模的征服运动，以掠夺的方式开发利用自然资源。为了满足农业生产和建设用地的需要，在很长的一段时间里，人们对湿地进行无度的开发利用，致使大量的湿地遭到破坏。据估计，我国有40%的重要湿地受到退化的威胁，历史上湿地资源丰富的长江中下游地区、三江平原、松嫩平原及若尔盖草原、东部沿海滩涂、河口三角洲及红树林等天然湿地区域都曾经历过严重丧失和"消亡"。

据统计，工业文明大力发展的20世纪，占世界人口15%的工业发达国家，消费了世界56%的石油、60%的天然气、50%的重要矿产资源。此外，包括湿地在内的自然生态系统的破坏带来的生物多样性下降、水资源危机、洪涝等极端灾害及一系列严重的生态、环境问题，引发了人类进一步地思考和担忧——要实现工业化，如果沿袭传统工业化的发展方式，是否还有余地和空间。人类社会第一次遇到了前所未有的生存与发展危机。在此背景下，人类开始了生存与发展的深刻反思和艰难探索，生态文明顺势而生。1972年在瑞典斯德哥尔摩召开的联合国人类环境会议，唤起了各国政府对环境问题的关注；1992年在巴西里约热内卢召开的联合国环境与发展大会，使可持续发展思想得到了最高级别的政治承诺，为生态文明建设提供了保障。

在这一时期，《湿地公约》作为世界上第一个旨在使全体国际社会成员普遍参加的自然保护公约，同时也是世界上唯一针对单一生态系统进行保护的国际条约

长江三峡　Yangtze River Three Gorges

丹江口水库　Danjiangkou Reservoir

正式签订。湿地作为从古至今与人类生存发展休戚相关的重要生态系统和生活环境，由此更被视为实现人类生态文明的重要载体。🈂

二、我国生态文明建设战略与湿地保护

1. 我国的生态文明建设战略

2007年10月召开的中国共产党第十七次全国代表大会上，首次提出"建设生态文明"，并将其列为全面建设小康社会的一项重要目标。这一重大命题的提出，标志着中国共产党发展理念的升华，对发展与环境关系认识的飞跃、治国方略的创新和发展，具有划时代的意义。而在此之前，我国就已经陆续开始了对"生态文明建设"的探索，在转变发展模式的思考中，全国生态省、生态市、生态县建设蓬勃兴起。

2012年11月召开的中国共产党第十八次全国代表大会上，明确提出生态文明建设应当放在突出地位，并将其纳入建设中国特色社会主义事业"五位一体"的建设布局中，强调在大力进行经济建设、政治建设、文化建设、社会建设的同时，增强和保证生态文明建设。这是对党的十七大以来生态文明理论研究成果以及我国生态省、生态市、生态县建设成就和经验的总结和再创造。自此，生态文明深刻融入和全面贯穿我国经济、政治、文化、社会建设的各个方面，成为我国发展战略的新思想，具有重要的现实意义，并将产生深远的历史影响。

2. 我国的湿地保护战略

党的十八大报告首次把"美丽中国"作为生态文明建设的宏伟目标，明确提出"加大自然生态系统和环境保护力度""要实施重大生态修复工程""扩大森林、湖泊、湿地面积，保护生物多样性"。党的十八大以来，党中央、国务院高度重视湿地保护工作，每年中央一号文件和国务院工作报告都对湿地保护提出了要求。2015年4月，《中共中央　国务院关于加快推进生态文明建设的意见》明确提出了"湿地面积不低于8亿亩"的目标，并进一步提出"自然岸线保有率不低于35%""构建平衡适宜的城乡建设空间体系，适当增加生活空间、生态用地，保护和扩大绿地、水域、湿地等生态空间""实施重大生态修复工程，扩大森林、湖泊、湿地面积""启动湿地生态效益补偿和退耕还湿。加强水生生物保护，开展重要水域增殖

碧水天镜　a blue water, a mirror

放流活动""研究建立江河湖泊生态水量保障机制""积极应对气候变化……增加森林、草原、湿地、海洋碳汇等手段""研究制定……湿地保护、生物多样性保护、土壤环境保护等方面的法律法规""对水流、森林、山岭、草原、荒地、滩涂等自然生态空间进行统一确权登记,明确国土空间的自然资源资产所有者、监管者及其责任""科学划定森林、草原、湿地、海洋等领域生态红线,严格自然生态空间征(占)用管理,有效遏制生态系统退化的趋势""加快推进……湿地……等的统计监测核算能力建设"。

2015年9月,中共中央、国务院印发《生态文明体制改革总体方案》进一步指出,"必须保护森林、草原、河流、湖泊、湿地、海洋等自然生态""中央政府主要对……大江大河大湖和跨境河流、生态功能重要的湿地草原、海域滩涂、珍稀野生动植物种和部分国家公园等直接行使所有权""开展水流和湿地产权确权试点。探索建立水权制度,开展水域、岸线等水生态空间确权试点,遵循水生态系统性、整体性原则,分清水资源所有权、使用权及使用量"。

为加快推进国家生态文明建设,全面贯彻落实党中央、国务院关于湿地保护的有关要求,2016年11月国务院办公厅印发《湿地保护修复制度方案》,明确提出我国湿地保护的任务目标,即实行湿地面积总量管控,到2020年,全国湿地面积

大九湖湿地晨雾　　morning fog of Dajiu Lake

张家湖湿地荷花　　*Nelumbo nucifera* in Zhangjia Lake wetland

不低于8亿亩，其中，自然湿地面积不低于7亿亩，新增湿地面积300万亩，湿地保护率提高到50%以上。严格湿地用途监管，确保湿地面积不减少，增强湿地生态功能，维护湿地生物多样性，全面提升湿地保护与修复水平。《湿地保护修复制度方案》是我国生态文明体制改革的重要成果，这是继2004年国务院办公厅下发《关于加强湿地保护管理的通知》做出"抢救性保护"重大举措以来，党中央、国务院关于湿地保护的最新顶层制度设计，开启了我国"全面保护湿地"的新篇章。

2017年10月，在第三次全国国土调查工作分类中，将湿地明确为一级地类，包括红树林地，天然的或人工的，永久的或间歇性的沼泽地、泥炭地、盐田、滩涂等。2017年11月，国家质检总局、国家标准化管理委员会批准发布实施《土地利用现状分类》（GB/T 21010—2017）国家标准。将"湿地"调整为与耕地、园地、林地、草地、水域等并列的一级地类，极大地提升了湿地资源的地位，也为全面适应生态文明建设、机构改革要求和加强湿地管理的需要奠定了坚实的基础。

2017年10月召开的中国共产党第十九次全国代表大会报告中要求"加大生态系统保护力度。……强化湿地保护和恢复"。党的十九大报告对生态文明建设浓墨重笔，"生态文明"被提及12次，"美丽"被提及8次，"绿色"被提及15次，更

丹江口库区　Danjiangkou Reservoir Area

首次提出建设富强民主文明和谐美丽的社会主义现代化强国的目标，提出现代化是人与自然和谐共生的现代化。强调人与自然是生命共同体，人类必须尊重自然、顺应自然、保护自然。这些理念把生态文明建设上升到了全新的高度，为美丽中国建设制定了时间表、路线图。

2018年，湿地保护立法作为第三类研究论证项目正式列入十三届全国人大常委会立法规划，并列入2020年度立法工作计划预备项目和2021年度立法工作计划。2021年12月24日，十三届全国人大常委会第三十二次会议表决通过了《中华人民共和国湿地保护法》，国家主席习近平同日签署第102号主席令并予以公布，自2022年6月1日起施行。这是我国首部专门针对湿地保护的法律，《湿地保护法》明确了长期以来引发部门管理争议的湿地概念，理顺了我国湿地的管理体制，通过加强湿地资源基础管理、发挥规划在湿地保护中的引领作用、完善湿地分级分类保护制度、科学推进湿地修复、强化湿地生态功能和生物多样性保护、规范引导湿地的合理利用、鼓励公众参与、加强科学研究、人才培养与国际合作等法律规定，从湿地生态系统的整体性和系统性出发，确立了湿地保护管理顶层设计的"四梁八柱"。随着《湿地保护法》的出台实施，我国的湿地保护迈入高质量发展的新阶段。

三、习近平总书记高度重视湿地保护

1. 习近平生态文明思想指导下的湿地保护

党的十八大以来，习近平总书记从我国生态建设实际出发，做出了一系列关于社会主义生态文明建设的重要论述，深刻回答了为什么建设生态文明、建设什么样的生态文明、怎样建设生态文明等重大理论和实践问题，形成了系统的习近平生态文明思想。2018年，习近平总书记在全国生态环境保护大会上强调：坚决打好污染防治攻坚战，推动生态文明建设迈上新台阶，并首次系统提出生态文明建设的六项原则。习近平生态文明思想是我国改革开放以来生态文明建设的最新成果，是习近平生态文明思想的重要精髓，也是推进我国生态文明建设迈上新台阶的根本遵循和行动指南。深刻理解和领悟习近平生态文明思想指导下的湿地保护工作，对全面推进我国湿地保护的高质量发展具有重要的意义。

（1）"坚持人与自然和谐共生"与湿地保护

人与自然和谐共生，是习近平生态文明思想的世界观和方法论基础，也是中

黄石网湖湿地　Wanghu Lake wetland in Huangshi

梁子湖湿地　Liangzi Lake wetland

利川香安坝　Xiang'an dam in Lichuan

国现代化的重要特征。党的十九大报告把"坚持人与自然和谐共生"作为新时代坚持和发展中国特色社会主义的基本方略之一。湿地作为地球三大生态系统之一，是人、自然、生物组成的综合体，是打造人与自然和谐共生的重要载体。

乡村振兴，是党的十九大提出的七大战略之一，是乡村生产、生活、生态的全面振兴。湿地保护是实现乡村振兴的重要"抓手"。湿地是美丽乡村、美好生活的重要标志，这为扩大湿地生态空间提供了广阔的天地。湿地作为古往今来实现人与自然和谐共生的重要载体，也必将为建设新时代人与自然和谐共生的现代化社会发挥重要作用。

（2）"绿水青山就是金山银山"与湿地保护

"绿水青山就是金山银山"的生态发展观是"坚持人与自然和谐共生"的基本构成，是从实际出发，把握经济发展与生态环境共存的重要科学论断。习近平"绿水青山就是金山银山"的论断与"既要绿水青山，也要金山银山""宁要绿水青山，不要金山银山"是辩证统一的。正如习近平总书记所说的："不要将生态环境保护和发展对立起来，要正确处理二者间的关系，既要加快发展又要守护好生态。"湿地作为地球上最重要的生态系统和人类生活依存最重要的自然环境之一，在发挥

绿水青山就是金山银山　　Lucid waters and lush mountains are invaluable assets

重要生态功能的同时，也具有巨大的经济效益和社会效益。《湿地公约》在全球范围推广的湿地保护理念强调的是湿地保护与合理利用，在保护的前提下开展生态旅游、休闲观光、生态种养等。做好湿地保护工作本身就是践行"绿水青山就是金山银山"的有力举措。

（3）"良好生态环境是最普惠的民生福祉"与湿地保护

习近平总书记指出："建设生态文明，是民意，也是民生""环境治理是一个系统工程，必须作为重大民生实事紧紧抓在手上""良好生态环境是最公平的公共产品，是最普惠的民生福祉"。习近平生态文明思想，体现了炽热的民生情怀。良好的生态环境是人类健康与幸福的必要条件。当前，我国湿地保护存在地区发展不平衡、优质湿地产品供应短缺等问题，远远不能满足人们对新鲜的空气、纯净的水、绿色的产品的需求。湿地工作是我国生态文明建设的短板，为解决新时代我国社会的主要矛盾，为人民群众提供更多湿地生态公共产品，提高生活质量和幸福指数，让老百姓在分享发展红利的同时，更充分地享受绿色福利，使生态文明建设成果更好地惠及全体人民，造福子孙后代，强化湿地保护和恢复将是一个重要的突破口。

（4）"山水林田湖草沙是生命共同体"与湿地保护

习近平生态文明思想的一个重要方面，是山水林田湖草沙的系统观、全局观和整体观。习近平总书记指出："山水林田湖是一个生命共同体，人的命脉在田，田的命脉在水，水的命脉在山，山的命脉在土，土的命脉在树。"山水林田湖草沙是生态系统内相互关联的构成要素，当它们之间能协调地发挥相互作用时就实现了人类社会发展的平衡。

山水林田湖草沙中的每一环都是生态文明建设的重要一环，从单一的角度来看，它们有各自的功能，但是从整体角度来看，生态环境离不开其中的任何一个部分。湿地是由各单项自然要素构成但又相互依存的统一综合体，是具有多种功能、各功能相互支撑、相互制约的完整生态系统，是"山水林田湖草沙生命共同体"中相对复杂和脆弱的一环。湿地作为生态系统，只要其中一个要素或一个环节出现问题，都将导致整个湿地资源综合体或生态系统的破坏甚至崩溃。同时，湿地生态系统类型众多，具有显著的时空特异性和综合性特征，这就要求在保护湿地的过程中，将湿地生态系统和各个生态要素视为一个整体，着力提高湿地生态系统自我修复能力，切实增强湿地生态系统的稳定性，显著提升湿地生态系统功能，

城市湿地
（武汉解放公园）
urban wetlands
(Wuhan Jiefang Park)

峡谷风情
valley landscape

钟祥客店镇赵泉河村
Zhaoquanhe village in Kedian
town of Zhongxiang City

大别山春晨　one spring morning in Dabie Mountains

为打造山水林田湖草沙生命共同体、维护国家生态安全、加快建设美丽中国奠定坚实生态基础。

（5)"用最严格制度最严密法治保护生态环境"与湿地保护

习近平总书记指出："只有实行最严格的制度、最严密的法治，才能为生态文明建设提供可靠保障。"与森林、草原、海洋三种类型自然生态系统相比，湿地的法制化保护进程时滞近40年。《湿地保护法》的出台，填补了我国生态系统立法空白，是健全完善我国生态文明制度体系的重要举措。

与我国现有的森林法、草原法等单一自然资源法相比较，《湿地保护法》更加注重生态系统的保护和修复，制度设计更多地从湿地生态系统整体性保护出发。法律中制度设计系统全面，设置了如部门协作机制、总量控制制度、调查评价制度、修复制度、约谈制度等法律制度，形成了湿地生态系统保护和修复制度的统一有机整体，有利于实现湿地保护高质量发展。

(6)"共谋全球生态文明建设"与湿地保护

"共谋全球生态文明建设"体现了习近平生态文明思想更为广阔、更为长远的蓝图,是对国际关系和全球治理的宏大构想。党的十九大报告把"坚持推动构建人类命运共同体"作为新时代中国特色社会主义的基本方略之一。我国湿地保护进程的不断推进,也反映了我国为国际湿地保护贡献"中国智慧"、提供"中国方案"的不断探索,在全方位保障我国全面履行《湿地公约》《生物多样性公约》《联合国气候变化框架公约》等国际环境公约的同时,更体现了我国长期致力于"身体力行"地推动构建人类命运共同体,谋求全球生态文明之路,为实现更高水平的全球可持续发展所做出的不懈努力。

2. 习近平总书记针对湿地保护的重要讲话

党的十八大以来,习近平总书记多次在有关会议和考察中强调湿地的重要性,高度重视湿地的保护和恢复工作。2013年12月,习近平总书记在中央城镇化工作会议上的讲话提出,要"扩大森林、湖泊、湿地等绿色生态空间比重,增强水源涵养能力和环境容量","稻田、麦浪、青纱帐、湿地都可以成为城市风景",针对城市缺水的现状,他指出,"为什么这么多城市缺水?一个重要原因是水泥地太多,把能够涵养水源的林地、草地、湖泊、湿地给占用了,切断了自然的水循环,

鄂州洋澜湖湿地公园　Yanglan Lake Wetland Park in Ezhou

雨水来了，只能当作污水排走，地下水越抽越少。解决城市缺水问题，必须顺应自然。"2014年3月，习近平总书记在中央财经领导小组第五次会议上的讲话中强调，"湖泊湿地是'地球之肾'"，针对我国湖泊湿地大量减少的状况，他提出"要采取硬措施，制止继续围垦占用湖泊湿地的行为，对有条件恢复的湖泊湿地要退耕还湖还湿。"2015年1月，习近平总书记在云南考察工作，在详细了解洱海湿地生态保护情况时强调，"要把生态环境保护放在更加突出位置，像保护眼睛一样保护生态环境，像对待生命一样对待生态环境，在生态环境保护上一定要算大账、算长远账、算整体账、算综合账，不能因小失大、顾此失彼、寅吃卯粮、急功近利。生态环境保护是一个长期任务，要久久为功。"习近平总书记在与当地干部合影后说："立此存照，过几年再来，希望水更干净清澈。"他叮嘱："一定要把洱海保护好，让'苍山不墨千秋画，洱海无弦万古琴'的自然美景永驻人间。"

2016年1月，习近平总书记在重庆召开推动长江经济带发展座谈会，听取对推动长江经济带发展的意见和建议并发表重要讲话，他在讲话中指出，"当前和今后相当长一个时期，要把修复长江生态环境摆在压倒性位置，共抓大保护，不搞大开发。要把实施重大生态修复工程作为推动长江经济带发展项目的优先选项。"2018年4月，习近平总书记实地考察长江经济带发展战略实施情况时，前往湖南省岳阳市君山华龙码头，察看取缔非法砂石码头及整治复绿、湿地修复情况时再次强调，"绝不容许长江生态环境在我们这一代人手上继续恶化下去，一定要给子孙后代留下一条清洁美丽的万里长江！"随后他在武汉主持召开深入推动长江经济带发展座谈会上提出，"抓湿地等重大生态修复工程时有没有先从生态系统整体性特别是从江湖关系的角度出发，从源头上查找原因，系统设计方案后再实施治理措施？"2019年8月，习近平总书记在甘肃考察时强调，"黄河、长江都是中华民族的母亲河，保护母亲河是事关中华民族伟大复兴和永续发展的千秋大计。"2020年1月，习近平总书记在昆明滇池星海半岛生态湿地考察调研时强调，"我们要避免走先污染再治理的老路，一定要摒弃过去那种以生态环境为代价换取一时经济发展的做法。我们提出新发展理念、建设生态文明，是符合客观规律的。从当前看，老百姓现在吃饱穿暖了，最关心的就是环境。长远来讲，我们不能吃子孙饭，要造福人类。要继续抓下去，锲而不舍、久久为功，把绿水青山真正变成金山银山。"2020年3月，习近平总书记在浙江杭州西溪国家湿地公园考察湿地保护利用情况时提出,"湿地贵在原生态,原生态是旅游的资本,发展旅游不能牺牲生态环境,不能搞过度商业化开发,不能搞一些影响生态环境的建筑,更不能搞私人会所,让公园成为人民群众共享的绿色空间。"2020年8月,习近平总书记就加强防汛救灾和灾后恢复重建等工作深入安徽考察调研。在详细了解巢湖防汛救灾和固坝巡堤查险工作后,习近平总书记强调,要坚持生态湿地蓄洪区的定位和规划,防止被侵占蚕食,保护好生态湿地的行蓄洪功能和生态保护功能。他强调,"洪水退后,要防止蓄洪区内出现水退人进的现象。我们要实现人与自然和谐相处,就不能同自然争夺发展空间。"2021年10月20日,习近平总书记来到山东东营黄河三角洲国家级自然保护区考察。习近平总书记强调,"要管理好,不能让湿地受到污染,也不能打猎、设网捕鸟""要把保护黄河口湿地作为一项崇高事业,让生态文明理念在实现第二个百年奋斗目标新征程上发扬光大,为实现社会主义现代化增光增色"。

石首麋鹿国家级自然保护区　Shishou Milu National Nature Reserve

习近平总书记关于湿地保护的重要论述，是习近平生态文明思想的重要组成部分，其精神实质和深刻内涵，也为湿地保护工作进一步指明了方向。🀄

四、我国湿地保护成效

我国于1992年正式向联合国教科文组织（UNESCO）递交了由时任中国外交部部长钱其琛签署的无任何保留条款加入书，同年7月31日加入书正式生效，成为《湿地公约》第67个缔约方。加入《湿地公约》30年来，在党和国家的高度重视下，在林草部门的牵头组织下，在各部门和全社会的大力支持及国际社会的帮助下，我国湿地保护取得显著成效。

1. 持续推进湿地保护修复工作

2003年，国务院批准了《全国湿地保护工程规划（2002—2030）》，将其作为今后一段时期开展湿地保护的指导性文件，国家组织实施了三个五年实施规划，党的十八大以来，中央政府累计投入168.55亿元，实施了3400多个湿地保护项目，新增和修复湿地80余万公顷。2016—2020年，中央投入98.7亿元，全国累计完成退化湿地修复（不含退耕还湿）面积30.45万公顷、退耕（退渔、退垸、退牧）还湿面积6.65万公顷、农作物等损失补偿面积87.71万公顷。湿地保护修复工作使退化湿地生态状况得到大幅改善，有力带动了各地方共同开展湿地保护与修复工作。

2. 不断健全湿地保护法规和制度体系

2016年11月国务院办公厅印发《湿地保护修复制度方案》，国家和省级层面陆续出台实施方案，开启了我国湿地的全面保护。国家林业和草原局先后出台了《湿地保护管理规定》，印发了《国家湿地公园管理办法》《国家重要湿地认定和名录发布规定》等文件，28个省（区、市）出台湿地保护条例等法规和规章制度，2021年12月24日，《中华人民共和国湿地保护法》正式出台，并于2022年6月1日起正式施行。《湿地保护法》坚持山水林田湖草沙是生命共同体的理念，从维护湿地生态系统整体性出发，建立湿地保护修复制度，增强湿地生态功能，维护湿地生物多样性，保障生态安全，促进生态文明建设。《湿地保护法》的颁布标志着我国湿地保护工作进入法治化的新阶段。

清江画廊　Qingjiang River Gallery

远安县嫘祖故里　Leizu hometown in Yuanan county

武汉东湖　East Lake in Wuhan

3. 不断完善湿地保护机构和体系建设

2018年党和国家机构改革以来，国家层面和各省级层面先后建立湿地保护机构，截至2022年8月，我国政府指定了国际重要湿地64处，认定国家重要湿地29处、省级重要湿地1001处，建立了三江源等5处国家公园、内陆湿地和水域生态系统类型国家级自然保护区69处、1693处湿地公园（其中国家湿地公园901处）和一大批湿地保护小区；常德、常熟、东营、哈尔滨、海口、银川6座城市入选全球首批国际湿地城市；2022年6月，武汉、盘锦、南昌、盐城、合肥、梁平、济宁7个城市获得国际湿地城市称号；全国湿地保护率达50%以上，湿地保护体系初步建立，湿地保护管理工作有序推进。

4. 稳步推动湿地调查监测体系建设

我国先后于1995—2003年和2009—2013年组织开展了两次全国湿地资源调查，摸清了全国湿地面积、分布、保护和利用等重要基础数据。每年对国际重要湿地进行生态状况监测，定期发布《国际重要湿地生态状况》白皮书，建立40处湿地生态定位站。与有关部门建立了泥炭地调查合作机制，已完成吉林等7省

沉湖湿地　Chenhu Lake wetland

（区）调查任务。并已将湿地监测纳入国家林草生态综合监测评价，不断夯实湿地保护基础。

5. 不断深化湿地履约和国际合作

我国认真履行《湿地公约》，深度参与公约事务和国际规则制定，推动中国湿地保护修复经验和模式形成公约决议，广泛开展国际交流合作，为全球生态治理贡献中国智慧和中国方案。2022年11月，《湿地公约》第十四届缔约方大会在我国湖北省武汉市举办，大会主题为"珍爱湿地 人与自然和谐共生"，充分体现了习近平生态文明思想和人与自然和谐共生理念，体现了我国长期致力于推进构建人类命运共同体、全面履行国际义务、承担大国责任、展现大国担当，为实现更高水平的全球可持续发展所做出的不懈努力。

虽然我国湿地保护取得了一系列显著成效，但我国湿地仍然面临严峻的问题和挑战。我国湿地率远低于全球8.6%的平均水平，人均湿地面积仅为世界人均面积的1/5。伴随社会经济的快速发展，湿地保护和利用的矛盾仍然存在，湿地遭受非法侵占、围垦、污染等现象时有发生。根据我国开展的两次全国湿地资源调查结果，相同口径下的湿地面积减少了339.63万公顷，威胁湿地生态状况主要因子由3个增加到了5个，威胁因子出现频次增加了38.72%，反映出我国湿地生态系统仍然面临很大威胁。一些地方还存在无序开发破坏湿地的行为，在长江经济带、沿海等区域尤为突出，这在一定程度上导致湿地防洪蓄洪能力大大降低，洪涝灾害发生的频次和强度显著增加，给人民生命财产造成重大损失。此外，还存在湿地保护修复系统性、科学性不够，湿地保护修复支撑体系薄弱等问题。总体上我国的湿地保护工作仍面临较大压力。

Chapter 1

Ecological Civilization and Wetland Conservation

Wetlands are an important part of the earth's ecosystem. Together with forests and oceans and known as the "kidney of the earth", they constitute the three major ecosystems of the earth and have irreplaceable system functions. Wetlands, as an important vehicle for the realization of human ecological civilization, can provide freshwater resources, food sources, purify the water environment, maintain ecological balance, and mitigate climate change, and are closely related to human survival, reproduction and development. The 18th National Congress of the CPC incorporated the construction of ecological civilization into the Five-sphere Integrated Plan of the socialism with Chinese characteristics, and wetland conservation is an important element of ecological civilization which has a direct bearing on national ecological security, sustainable economic and social development, and the survival and well-being of future generations of the Chinese nation. China's wetland conservation has gone through three stages: investigating the status quo thoroughly and consolidating foundation (1992-2003), rescuing conservation (2004-2015), and comprehensive conservation (2016-2021). *The Wetland Conservation Law of the People's Republic of China* came into force on June 1, 2022. This is an important legislative achievement to implement the decisions and deployment of the Party Central Committee, practice Xi Jinping's thought of ecological civilization, and further strengthen the conservation and restoration of wetland ecosystems, which also indicates that China's wetland conservation will enter into a new phase of quality development in a new era.

1.1 Wetlands, the vehicle for human ecological civilization

Wetlands are the most productive ecosystems on earth. It is estimated that the value of ecosystem services provided by global wetland ecosystems is as high as $47 trillion per year, higher than the sum of the value of global forest, desert and grassland ecosystem services. In order to protect wetlands and wetland resources worldwide, representatives from 18 countries signed the Convention of Wetlands of International Importance Especially as Waterfowl Habitats (the *Wetlands Convention*, also known as the *Ramsar Convention*) in the seaside town of Ramsar, Iran, on February 2, 1971, aiming to protect and wisely use wetlands through local and national actions and international cooperation, and to contribute to sustainable development throughout the world.

Wetlands are defined in the *Ramsar Convention* as "areas of marsh, fen, peatland or water, whether natural or artificial, permanent or temporary, with water that is static or flowing, fresh, brackish or salt, including areas of marine water the depth of which at low tide does not exceed six meters". According to the classification system of the *Ramsar Convention*, wetlands worldwide can be broadly classified as natural wetlands and artificial wetlands. Natural wetlands can be further divided into offshore and coastal wetlands as well as inland wetlands. Coral reefs, seagrass beds, mangroves, tidal flats, salt marshes in the coastal areas are representatives of offshore and coastal wetlands; lakes, rivers, freshwater marshes, peatlands are typical inland wetlands. Common artificial wetlands include fish ponds, salt pans, reservoirs, and rice paddies.

Throughout history, wetlands have been closely related to the development of human civilization. The four ancient civilizations of Egypt, Babylon, India and China, which are famous in human history, were all nurtured and flourished in wetlands. Many of the world's river wetlands are the "cradle" of the original human civilization. In an agricultural society where the human productivity level is low, people usually choose an area with suitable climate, abundant water, and fertile land to cultivate, live and establish settlements. In ancient China, the Yellow River and Yangtze River nourished the Chinese prople, and our ancestors created the ancient and splendid Chinese civilization by these two rivers. The abundant water resources of the Yellow River and Yangtze River and the fertile land on their both sides provided the prerequisites for the development of agriculture and animal husbandry, and the history of Chinese civilization thus originated along with the wetlands and grew slowly along with the sound of murmuring water, nourishing and prolonged.

After the development of primitive and agrarian civilization, along with the industrial revolution, mankind ushered in industrial civilization, in which mankind launched an unprecedented conquest campaign against nature, exploiting natural resources in a plundering manner. In order to meet the needs of agricultural production and construction land, for a long period of time, people exploited wetlands in an inordinate manner, resulting in the destruction of a large number of wetlands. It is estimated that 40% of China's important wetlands are threatened by serious degradation, and the middle and lower reaches of the Yangtze River, Sanjiang Plain, Songnen Plain and Ruoergai Prairie, tidal flats along the eastern coast, estuarine deltas and mangrove forests, which are historically rich in wetland resources, have suffered serious loss of natural wetlands and even extinction.

According to statistics, throughout the 20th century when the industrial civilization thrived, industrially developed countries, accounting for 15% of the world's population, consumed 56% of the world's oil and 60% of natural gas and 50% of the important mineral resources. In addition, the destruction of wetlands and other natural ecosystems has brought about a series of serious ecological and environmental problems, such as the decline of biodiversity, the water crisis, and extreme climatic disasters such as floods, which have also triggered further thinking and concern among human beings about whether there is still room and space for industrialization if we follow the traditional industrialized way of development. For the first time, human society has encountered an unprecedented crisis of survival and development. The United Nations Conference on Human Environment held in 1972 aroused the attention of governments to environmental issues, and the United Nations Conference on the Environment and Development held in Rio de Janeiro, Brazil, in 1992 brought the idea of sustainable development to the highest level of political commitment and provided a guarantee for the construction of ecological civilization.

During this period, the *Ramsar Convention* was officially signed as the world's first nature conservation convention aimed at the universal participation of all members of the international community, and the only international treaty in the world to deal with the conservation of a particular ecosystem. As an important ecosystem and living environment that has been closely related to human survival and development since ancient times, wetlands are regarded as an important vehicle for the realization of human ecological civilization.

1.2 China's construction strategy of ecological civilization and wetland conservation

1.2.1 China's construction strategy of ecological civilization

At the 17th National Congress of the Communist Party of China held in October 2007, "building an ecological civilization" was proposed for the first time and listed as an important goal for building a moderately prosperous society in all aspects. The introduction of this important proposition marked the sublimation of the development concept of the CPC, the leap in the understanding of the relationship between development and environment, and the innovation and development of the strategy of governance, which is of epoch-making significance. Before that, China had already started to explore the construction of ecological civilization, and the construction of eco-provinces, eco-cities and eco-counties had flourished in the thinking of changing the development mode.

At the 18th National Congress of the Communist Party of China held in November 2012, it was clearly proposed that the construction of ecological civilization should be put in a prominent position and incorporated into the construction layout of Five-sphere Integrated Plan for building socialism with Chinese characteristics, emphasizing that while vigorously carrying out economic, political, cultural and social construction, the construction of ecological civilization should be enhanced and guaranteed. This is a summary and re-creation of the theoretical research results of ecological civilization since the 17th National Congress, as well as the achievements and experience of eco-provinces, eco-cities and eco-counties in China. Since then, ecological civilization has been profoundly integrated into and comprehensively penetrated into all aspects of China's economic, political, cultural and social construction, becoming a new idea of our national development strategy, which has important practical significance and will have far-reaching historical impact.

1.2.2 China's wetland conservation strategy

The report of the 18th National Congress put "beautiful China" as the grand goal of construction of ecological civilization for the first time, and clearly proposed "to increase the efforts at conservation of natural ecosystem and environment, implement major ecological restoration projects, expand the area of forests, lakes and wetlands, and protect biodiversity". Since the 18th National Congress, the CPC Central Committee and the State Council have attached great importance to wetland conservation, and each year No. 1 central document and the report on the work of the government have put forward requirements for wetland conservation. In April 2015, the Opinions of the CPC Central Committee and the State Council on Accelerating the Construction of Ecological Civilization clearly put forward the goal of "wetland area of not less than 800 million mu", and further proposed that "the retention rate of natural shoreline is not less than 35%", "to build a balanced and appropriate urban and rural construction space system, appropriately increase living space, ecological land, and protect and expand green space, waters, wetlands and other ecological space", "to implement major ecological restoration projects, expand the area of lakes and wetlands", "to start compensation of wetland ecological benefits and return land for farming to wetlands, strengthen the conservation of aquatic organisms, and carry out fish stockings in important waters", "to study the establishment of a mechanism to guarantee ecological water quantity in rivers and lakes", "to actively respond to climate change and increase the means of forests, grasslands, wetlands and ocean carbon sinks", "to research and formulate laws and regulations on wetland conservation", "to clarify the owners and supervisors of natural resource assets in national land space and their responsibilities through unified registration of rights of natural ecological space such as water flows and tidal flats" "to scientifically delineate ecological red lines in wetlands and other areas, strictly managing the requisition (occupation) of natural ecological space and effectively curbing the trend of ecosystem degradation", and "to accelerate the construction of statistical monitoring and accounting capacity for wetlands".

In September 2015, the CPC Central Committee and the State Council issued the Integrated Reform Plan for Promoting Ecological Civilization further pointed out that "It is necessary to protect the natural ecology such as rivers, lakes, wetlands, etc.", "The central government will mainly exercise direct ownership over large rivers, lakes and cross-border rivers, wetlands and grasslands with important ecological functions, sea areas and tidal flats, rare wildlife species and some national parks", and "We should carry out pilot programes of confirmation of property rights of water flow and wetlands, explore the establishment of a water right system, carry out pilot right confirmation of waters, shorelines and other water ecological space, follow the principle of systemic and holistic water ecology, distinguish ownership of water resources from the right to use and the amount of use."

In order to accelerate the construction of national ecological civilization and fully implement the relevant requirements of the Central Committee of the Party and the State Council on wetland conservation, the General Office of the State Council issued the Wetland Conservation and Restoration System Scheme in November 2016, which clearly puts forward the task objectives of wetland conservation in China, namely, to control total wetland area, and by 2020, the national wetland area will not be less than 800 million mu with the natural wetland area no less than 700 million mu and the new wetland area of 3 million mu, increasing the wetland conservation rate to more than 50%. According to the Scheme, use of wetlands should be strictly supervised to ensure that the wetland area is not reduced, enhance the ecological functions of wetlands, maintain wetlands biodiversity, and comprehensively improve the level of wetland conservation and restoration. The issuance of the Wetland Conservation and Restoration System Scheme is an important achievement of the system reform of China's ecological civilization, which is the latest top-level system design of the Central Committee of the Party and the State Council on wetland protection since 2004, when the General Office of the State Council issued the Notice on Strengthening Wetland Conservation and Management to make a major initiative of rescuing conservation of wetlands, and opens a new chapter of "comprehensive wetland conservation" in China.

In October 2017, in the classification of the third national land survey launched, wetlands were classified as first-class land category, including mangroves, natural or artificial, permanent or intermittent marshes, peatlands, salt pans, tidal flats, etc. In November 2017, General Administration of Quality Supervision, Inspection and Quarantine and the Standardization Administration approved the release and implementation of the national standard Current Land Use Classification (GB/T 21010-2017). The adjustment of wetlands to a first-class land category alongside with arable land, garden land, forest land, grassland, waters, etc. has greatly elevated the status of wetland resources and laid a solid foundation to fully adapt to the construction of ecological civilization, the requirements of institutional reform and the need to strengthen wetland management.

The report of the 19th National Congress of the Communist Party of China held in October 2017 calls for "increasing the conservation of ecosystems, ..., and strengthening wetland conservation and restoration". It is heavily devoted to the construction of ecological civilization; "ecological civilization" is mentioned 12 times, "beautiful" 8 times, "green" 15 times, and the goal of building a rich, democratic, civilized, harmonious and beautiful socialist modernization power is proposed for the first time, and modernization is a modernization in which people and nature live together in harmony. It emphasizes that man and nature are a community of life; mankind must respect nature, conform to nature and protect nature. These ideas elevate the construction of ecological civilization to a whole new level, and set a timetable and roadmap for the construction of a beautiful China.

In 2018, wetland conservation legislation was formally included in the legislative planning of the Standing Committee of the 13th National People's Congress (NPC) as the third category of research and demonstration projects, and included in the preparatory items of the 2020 legislative work plan and the 2021 legislative work plan. On December 24, 2021, at the 32nd session of the 13th NPC, the Standing Committee voted to adopt the *Wetland Conservation Law of the People's Republic of China* (hereinafter referred to as the *Wetland Conservation Law*) and President Xi Jinping signed Presidential Decree No. 102 at the same day to promulgate it, effective from June 1, 2022. This is China's first law specifically for the wetland conservation. It clarifies the concept of wetland that has long led to sectoral management disputes, rationalizes the management system of China's wetlands, and establishes the "four beams and eight pillars" of the top-level design of wetland conservation and management from the holistic and systematic nature of the wetland ecosystem by strengthening the basic management of wetland resources, playing a leading role of planning in wetland conservation, improving the graded and classified wetland conservation system, scientifically promoting wetland restoration, strengthening ecological functions of wetlands and biodiversity conservation, regulating and guiding the rational use of wetlands, encouraging public participation, and strengthening scientific research, talent training and international cooperation and other legal provisions. With the introduction and implementation of the *Wetland Conservation Law*, China's wetland conservation will enter a new phase of quality development in a new era.

1.3 General Secretary Xi Jinping attaches great importance to wetland conservation

1.3.1 Wetland conservation under the guidance of Xi Jinping's thought of ecological civilization

Since the 18th NPC, General Secretary Xi Jinping has made a series of important expositions on the construction of socialist ecological civilization based on the actual ecological construction of China, profoundly answering the major theoretical and practical questions of why to build ecological civilization, what kind of ecological civilization to build, and how to build ecological civilization, and systematically formed the Xi Jinping's thought of ecological civilization. In 2018, General Secretary Xi Jinping stressed at China Eco-Environmental Protection Conference that we should resolutely fight the battle of pollution prevention and control, advance the construction of ecological civilization to a new level, and systematically put forward six principles of construction of ecological civilization for the first time. The six principles of Xi Jinping's thought of ecological civilization are the latest achievements in the construction of ecological civilization since China's reform and opening up, the important essence of Xi Jinping's thought of ecological civilization, and the fundamental guidance and action guide for advancing the construction of ecological civilization in China to a new level. A deep understanding and comprehension of wetland conservation under the guidance of the six principles of Xi Jinping's thought of ecological civilization is of great significance to comprehensively promote the quality development of wetland conservation in China.

1.3.1.1 "Ensure harmony between humanity and nature." and wetland conservation

The harmony between humanity and nature is the worldview and methodological basis of Xi Jinping's thought of ecological civilization, and an important feature of China's modernization. In the report of the 19th National Congress, "ensuring harmony between humanity and nature" is one of the basic policies for upholding and developing socialism with Chinese characteristics in the new era; and it proposed that the modernization we desire is a modernization in which man and nature, man and society live in harmony. As one of the three major ecosystems on the earth, wetlands are an integrated body composed of human, nature and organisms, and an important vehicle for building harmony between humanity and nature.

Rural revitalization, as one of the seven strategies proposed at the 19th National Congress of the CPC, is the comprehensive revitalization of rural production, life and ecology. Wetland conservation is an important "enabler" to realize rural revitalization. Wetlands are an important symbol of beautiful countryside and good life, which provide a lot of room for expanding ecological space for wetlands. As an important vehicle for the harmony between humanity and nature in the past and the present, wetlands will certainly play an important role in building a modern society in which man and nature live harmoniously in the new era.

1.3.1.2 "Lucid water and lush mountains are invaluable assets." and wetland conservation

The ecological development concept of "lucid water and lush mountains are invaluable assets", is the basic composition of "ensuring harmony between humanity and nature", which is an important scientific assertion that grasps the coexistence of economic development and ecological environment from the practical point of view. Xi Jinping's theory of "lucid water and lush mountains are invaluable assets" is a dialectical unification of the theory of "both lucid water and lush mountains and invaluable assets are important"and the theory of "lucid water and lush mountains are preferred". As General Secretary Xi Jinping said, "Instead of putting eco-environmental conservation and development in opposition to each other, we should correctly handle the relationship between the two, speeding up development while guarding the ecology."As one of the most important ecosystems on earth and the most important natural environments on which human life depends, wetlands play an important ecological function while also having huge economic and social benefits. The concept of wetland conservation promoted globally by the *Ramsar Convention* emphasizes the conservation and wise use of wetlands and the development of ecotourism, leisure and sightseeing, ecological breeding under the premise of conservation. It can be said that doing a good job of wetland conservation itself is a powerful measure to practice"lucid water and lush mountains are invaluable assets" .

1.3.1.3 "A good ecological environment is the most inclusive welfare of people." and wetland conservation

General Secretary Xi Jinping pointed out that "Building an ecological civilization is a matter of public opinion and people's livelihood", "Environmental management is a systematic project and must be held tightly as a major practical matter of people's livelihood", and"A good ecological environment is the fairest public product and the most inclusive welfare of people". Xi Jinping's thought of ecological civilization reflects the passionate sentiment towards people's livelihood. A good ecological environment is a necessary condition for human health and happiness. Currently China's wetland conservation has problems such as unbalanced regional development and shortage supply of quality wetland products, far from meeting people's demand for fresh air, pure water and green products. Wetlands are a weak point of China's construction of ecological civilization. In order to solve the main contradiction of our society in the new era, provide more public products of wetland ecology for the people, improve the quality of life and happiness index, allow the people to share the development dividend while enjoying the green welfare more fully, so that the construction achievements of ecological civilization can better benefit all people and their future generations, strengthening wetland conservation and restoration will be an important breakthrough.

1.3.1.4 "Mountains, rivers, forests, farmlands, lakes, and grasslands are a life community." and wetland conservation

An important aspect of Xi Jinping's thought of ecological civilization is the systemic, overall and holistic view of mountains, rivers, forests, farmlands, lakes, and grasslands. General Secretary Xi Jinping pointed out that "Mountains, rivers, forests, farmlands, lakes, and grasslands are a life community. The lifeblood of man is in the farmlands; the lifeblood of the farmlands is in rivers; the lifeblood of the rivers is in the mountains; the lifeblood of the mountains is in the earth; the lifeblood of the earth is in the trees." Mountains, rivers, forests, farmlands, lakes, and grasslands are interrelated components of the ecosystem; when they can play a coordinated role, a balance in the development of human society can be achieved.

From a single perspective, every link in the mountains, rivers, forests, farmlands, lakes, and grasslands is an important link in the construction of ecological civilization and they have their own functions, but from a holistic perspective, the ecological environment cannot be separated from any one of them. Wetlands are an unified complex composed of individual natural elements that are also interdependent. They are a complete ecosystem with a variety of functions, each of which is mutually supportive and mutually constrained, and also a relatively complex and fragile part of the "life community of mountains, rivers, forests, farmlands, lakes, and grasslands". As long as one of the elements or a link of the wetland ecosystem fails, the destruction or even collapse of the entire wetland resource complex or ecosystem will follow. At the same time, there are many types of wetland ecosystems, resulting in significant spatial and temporal specificity and comprehensive features, which requires that in the process of wetland conservation, we consider wetland ecosystems and various ecological elements as a whole, focus on improving the ability of wetland ecosystems to restore themselves, effectively enhance the stability of wetland ecosystems, significantly upgrade the function of wetland ecosystems in order to lay a solid ecological foundation for creating a life community of mountains, rivers, forests, farmlands, lakes, and grasslands, maintaining national ecological security and accelerating the construction of a beautiful China.

1.3.1.5 "Conserve the ecological environment with the strictest system and the most stringent rule of law." and wetland conservation

General Secretary Xi Jinping pointed out that "Only by implementing the strictest system and the most stringent rule of law can we provide a reliable guarantee for the construction of ecological civilization." Compared with the three types of natural ecosystems, namely forests, grasslands and oceans, the process of legalizing the wetland conservation has lagged behind for nearly four decades. The introduction of the *Wetland Conservation Law*, which fills the gap in China's ecosystem legislation, is an important initiative to improve and perfect the system of ecological civilization in China.

Compared with China's existing forest law, grassland law and other single natural resources law, *Wetland Conservation Law* focuses more on the conservation and restoration of ecosystems, and its system is designed more from the overall conservation of wetland ecosystems. The system design in the *Law* is systematic and comprehensive, setting up legal systems such as mechanisms of departmental collaboration, systems of total control, system of investigation and evaluation systems, restoration systems, interview systems, etc., forming a unified organic whole of conservation and restoration systems of wetland ecosystem, which is conducive to achieving quality development of wetland conservation.

1.3.1.6 "Seek to build a global ecological civilization." and wetland conservation

"Seek to build a global ecological civilization." reflects the broader and longer-term blueprint of Xi Jinping's thought of ecological civilization, and is a grand vision of international relations and global governance. The report of the 19th National Congress of the CPC has made "insisting on promoting the construction of a community of common destiny for all mankind" one of the ideas and basic policies of socialism with Chinese characteristics in the new era. The continuous advancement of China's wetland conservation process also reflects China's continuous exploration to contribute "China's wisdom" and provide "China's solutions" to international wetland conservation, which ensures China's full implementation of the *Ramsar Convention*, the *Convention on Biological Diversity* and the *United Nations Framework Convention on Climate Change*. It also reflects China's long-term commitment to promoting the construction of a community of common destiny for all mankind, seeking the path of global ecological civilization, and making unremitting efforts to achieve a higher level of global sustainable development.

1.3.2 General Secretary Xi Jinping's important speeches on wetland conservation

Since the 18th National Congress, General Secretary Xi Jinping has been repeatedly emphasizing the importance of wetlands in relevant meetings and visits, and attaching great importance to the conservation and restoration of wetlands. In December 2013, General Secretary Xi Jinping proposed in his speech at the Central Urbanization Work Conference that we should "expand the proportion of green ecological space such as forests, lakes and wetlands, and enhance the water conservation capacity and environmental capacity", "rice paddiees, wheat waves, green crops, wetlands can become urban scenery"; for the current situation of urban water shortage, he pointed out that "Why so many cities lack water? An important reason is that there are too many cement, the forest lands, grasslands, lakes, wetlands that can contain water are occupied, cutting off the natural water cycle, and when rainwater comes, it can only be treated as sewage discharge, and groundwater is pumped less and leess. To solve the problem of urban water shortage, we must follow nature." In March 2014, General Secretary Xi Jinping emphasized in his speech at the 5th session of the Central Leading Group for Financial and Economic Affairs that "Lakes and wetlands are the 'kidneys of the earth'", and in response to the massive reduction of lakes and wetlands in China, he proposed that "tough measures should be taken to stop the continued reclamation and occupation of lakes and wetlands, and to return land for farming to lakes and wetlands that are in a position to recover." In January 2015, General Secretary Xi Jinping inspected the work in Yunnan, after a detailed understanding of the ecological conservation of the wetlands of the Erhai, he stressed that "We must put ecological environmental protection in a more prominent position, protect the ecological environment as much as our eyes, treat the ecological environment as much as our lives, and must calculate the big account, the long-term account, the overall account, and the comprehensive account in ecological environmental protection, not to suffer a big loss for a little gain, attend to one thing and lose another, to eat next year's food, and to seek quick profits. Eco-environmental conservation is a long-term task that must be done over time." After taking a group photo with the local cadres, he said, "Keep this picture as a record and I will come back in a few years, hoping that the water will be cleaner and clearer." He urged, "Be sure to protect the Erhai well, so that the natural beauty of 'the Cang Mountain brightens without ink, the Erhai echoes without strings' lives on."

In January 2016, General Secretary Xi Jinping held a forum in Chongqing to promote the development of the Yangtze River Economic Belt and listen to the views and suggestions on promoting the development of the Yangtze River Economic Belt. He delivered an important speech, in which he pointed out that "At present and for quite a long period in the future, we must put restoration of the ecological environment of the Yangtze River in an overriding position, and promote well-coordinated environmental conservation and avoid excessive development. The implementation of major ecological restoration projects should be made a priority option in promoting the development projects of the Yangtze River Economic Belt." In April 2018, when General Secretary Xi Jinping went for a field inspection of the Yangtze River Economic Belt development strategy, he went to Junshan Huarong wharf, Yueyang, Hunan Province to inspect the banning of illegal sand and gravel wharf and remediation of re-greening as well as wetland restoration, and once again he stressed, "Never allow the ecological environment of the Yangtze River continue to deteriorate in the hands of our generation; we must leave a clean and beautiful Yangtze River to future generations!" Then he chaired a forum in Wuhan to promote the development of the Yangtze River Economic Belt in depth, and pointed out that "When we carry out wetland and other major ecological restoration projects, have we started from the perspective of the ecosystem as a whole, especially from the perspective of the relationship between rivers and lakes, and tried to find the cause from the source, or designed the program systematically before implementing governance measures?" In August 2019, General Secretary Xi Jinping stressed during a visit to Gansu, "The Yellow River and the Yangtze River are both the mother rivers of the Chinese nation. Protecting the Mother River is a matter of great rejuvenation and sustainable development of the Chinese nation." In January 2020, General Secretary Xi Jinping stressed during an inspection and research in the ecological wetlands of the Xinghai Peninsula in Dianchi Kunming that "We must avoid the old path of pollution before treatment, and must abandon the past practice of exchanging the ecological environment for one-time economic development. The new development concept and the construction of ecological civilization is in line with the objective laws. It is safe to say that the people are now fed and clothed, and they are most concerned about the environment. In the long run, instead of consuming the environmental resources of our children and grandchildren in advance, we must benefit mankind by continuing to focus on the task with perseverant and long-term efforts, so that the lucid waters and lush mountains really become the invaluable assets." In March 2020, General Secretary Xi Jinping went to the Xixi National Wetland Park in Hangzhou, Zhejiang Province to inspect the conservation and utilization of wetlands, and proposed that "Wetlands are valuable in their original ecology, which is the capital of tourism. The development of tourism can not sacrifice the ecological environment, or engage in excessive commercial development and buildings that affect the ecological environment, not to mention private clubs, so that the park becomes the green space people can share." In August 2020, General Secretary Xi Jinping made an in-depth investigation and research on strengthening flood control and disaster relief and post-disaster recovery and reconstruction in Anhui Province, after having a detailed understanding of the Chaohu Lake flood control and relief, dam consolidation and patrol for danger examination, Xi Jinping stressed the need to adhere to the positioning and planning of flood storage areas of ecological wetlands to prevent encroachment and protect the function of flood storage of ecological wetlands and function of ecological conservation. He stressed that "After the flood recedes, we should prevent the phenomenon of water receding and people advancing in the flood storage area. If we want to achieve harmony between man and nature, we cannot compete with nature for development space." On October 20, 2021, General Secretary Xi Jinping came to Yellow River Delta National Nature Reserve in Dongying, Shandong Province for inspection. General Secretary Xi stressed that "We must manage it well, and keep the wetlands from pollution. Hunting or setting up nets to catch birds should not be allowed", and "We must take the conservation of the Yellow River estuary wetlands as a noble cause, so that the concept of ecological civilization can flourish on the new journey to achieve the second century goal, and add light to the realization of socialist modernization." General Secretary Xi Jinping's important expositions on wetland conservation are an important part of Xi Jinping's thought of ecological civilization, and their spiritual essence and profound connotation have also further pointed out the direction for wetland conservation.

1.4 Progress of wetland conservation in China

On January 3, 1992, China formally submitted to UNESCO, the depositary of the *Wetland Convention*, the instrument of accession without any reservation signed by Qian Qichen, the then Chinese Minister of Foreign Affairs, and on July 31 of the same year, the instrument of accession formally entered into force, making China the 67th party to the *Wetland Convention* on. In the past 30 years since the accession to the *Wetland Convention*, with the high priority of the Party and the State, the leading organization of the forestry and grass sector, the strong support of other sectors and the whole society and the help of the international community, China has achieved remarkable results in wetland conservation.

1.4.1 Continuously improve wetland conservation and restoration

In 2003, the State Council approved the *National Wetland Protection Project Plan (2002-2030)* as a guiding document for carrying out wetland conservation in the future, and the state organized and implemented three five-year implementation plans. Since the 18th National Congress of CPC, the central government has invested a total of 16.855 billion yuan to implement more than 3,400 wetland conservation projects, adding and restoring more than 800,000 hectares of wetlands. From 2016 to 2020, the central government invested 9.87 billion yuan to complete the restoration of degraded wetlands (excluding the returning land for farming to wetlands) in a total area of 304,500 hectares, the returning land for farming (fishing, levee and grazing) to wetlands in an area of 66,500 hectares, and the compensation of crop and other losses in an area of 877,100 hectares. Wetland conservation and restoration has significantly improved the ecological condition of degraded wetlands, becoming a strong impetus for all local departments to work together to carry out wetland conservation and restoration.

1.4.2 Continuously perfect wetland conservation regulations and institutioanl systems

In November 2016, the General Office of the State Council issued the *Wetland Conservation and Restoration System Sheme*, and its implementation plans were issued one after another at the national and provincial levels, initiating the comprehensive conservation of wetlands in China. The National Forestry and Grassland Administration has issued the *Regulations on Management of Wetland Conservation*, *Management Measures of National Wetland Parks*, and *Identification and List Issuance Regulations on Wetlands of National Importance* and other documents, and 28 provinces, autonomous regions and municipalities have issued regulations and rules on wetland conservation. On December 24, 2021, the *Wet-

land Conservation Law of the People's Republic of China (*Wetland Conservation Law*) was officially introduced and came into force on June 1, 2022. The *Wetland Conservation Law* adheres to the concept that mountains, rivers, forests, farmlands, lakes and grasslands are a life community, and states that we should start from maintaining the integrity of wetland ecosystems, establish a conservation and restoration system of wetlands, enhance the ecological functions of wetlands, maintain wetland biodiversity to ensure ecological security and promote the construction of ecological civilization. The promulgation of the *Wetland Conservation Law* marks a new stage in the rule of law for the conservation of wetlands in China.

1.4.3 Continuously promote construction of wetland conservation institutions and systems

Since the reform of the CPC and state institutions in 2018, wetland conservation institutions have been established at the national and provincial level, and as of August 2022, our government has designated 64 wetlands of international importance, identified 29 wetlands of national importance and 1,001 wetlands of provincial importance, established 5 national parks including Sanjiangyuan, 69 national nature reserves for inland wetlands and water ecosystem types, and 1,693 wetland parks (including 901 national wetland parks) and a large number of smaller wetland conservation areas; six cities, namely Changde, Changshu, Dongying, Harbin, Haikou and Yinchuan have been selected as the first batch of international wetland cities in the world; in June 2022, seven cities, namely Wuhan, Panjin, Nanchang, Yancheng, Hefei, Liangping and Jining have been awarded the title of international wetland cities; wetland conservation rate has reached more than 50%. The national wetland conservation system has been basically established, allowing wetland conservation and management to be conducted in an orderly manner.

1.4.4 Steadily advance the construction of wetland survey and monitoring system

China has organized two national wetland resource surveys from 1995-2003 and 2009-2013, which have clarified important basic data on the area, distribution, conservation and utilization of wetlands nationwide. Every year, we monitor the ecological status of wetlands of international importance, regularly publish the white paper Ecological Status of Wetlands of International Importance, and establish 40 ecological positioning stations of wetlands. A peatland survey cooperation mechanism has been established with relevant departments, and survey tasks have been completed in seven provinces (regions), including Jilin Province. Wetlands monitoring has also been incorporated into the national comprehensive monitoring and evaluation of forestry and grassland ecology, continuously consolidating the foundation of wetland conservation.

1.4.5 Continuously deepen wetland compliance and international cooperation

China earnestly fulfills the *Wetland Convention*, deeply participates in the Convention and international rule-making, promotes China's experience and models of wetland conservation and restoration to form the Convention's resolutions, extensively carries out international exchanges and cooperation, and contributes China's wisdom and China's solutions to global ecological governance. In November 2022, the 14th Conference of the Contracting Parties to the Wetland Convention will be held in Wuhan, Hubei Province, China, with the theme of Cherish Wetlands, Harmony Between Man and Nature, fully reflecting Xi Jinping's thought of ecological civilization and the concept of harmony between man and nature, and China's long-term commitment to promoting the building of a common community of human destiny, fully fulfilling its international obligations, assuming the responsibilities of a great power, showing its role as a great power, and making the unremitting efforts to realize high-level global sustainable development.

Although China has achieved a series of remarkable results in wetland conservation, China's wetlands still face serious problems and challenges. China's wetland rate is far below the global average of 8.6%, and the per capita wetland area is only 1/5 of the world's per capita. Along with rapid socio-economic development, the contradiction between wetland conservation and utilization still exists; wetlands suffer from illegal encroachment, reclamation, pollution from time to time. According to the results of our two national wetland resources surveys, the area of wetlands under the same caliber decreased by 3,396,300 hectares; the main factors threatening the ecological status of wetlands increased from three to five; and the frequency of threatening factors increased by 38.72%, reflecting that China's wetland ecosystem is still confronted great threats. In some places, there is also disorderly development and construction that damage wetlands, especially in the Yangtze River Economic Belt and coastal areas, which to a certain extent has led to a significant reduction in the flood control and storage capacity of wetlands, and a notable increase of frequency and intensity of flooding, causing significant losses to people's lives and property. In addition, there are also problems such as the lack of systematic and scientific nature of wetland and restoration, and weak support system for wetland conservation and restoration. Overall China's wetland conservation is still under enormous pressure.

第二篇

长江大保护背景下
湖北湿地保护实践

CHAPTER 2

*HUBEI WETLAND CONSERVATION PRACTICES
IN THE CONTEXT OF YANGTZE RIVER
COORDINATED CONSERVATION*

一、湖北省湿地环境演变与资源

1. 湖北湿地形成与演变

湖北省湿地以长江、汉江为骨架，群带交错，湖库交融，构成完整多样的湿地生态系统，维系着长江流域乃至全国的湿地生态安全和生物多样性。

湖北湿地生态系统演变与长江有着密切的联系。史前时期，江汉平原是一个完整的湿地生态系统。晚更新世末至全新世初，湖北湿地呈现河湖交错的湿地景观，湿地面积较小，主要湿地类型为河流湿地和湖泊湿地，湿地以自然演替为主。

春秋战国时期，湖北湿地主要分布在江汉平原，为平原湖沼湿地景观，分布着"方九百里"的湖泊沼泽，云梦泽形成于江汉平原湖区。当时的云梦泽生物多样性丰富，湿地动植物随处可见，荆江有夏水、涌水等支流分流汇入云梦泽，并形成向东延展的陆上三角洲。春秋后期，楚国开凿废弃的分流道，连接汉江，这就是人工运河——杨水，而此时的下荆江还是湖沼区。秦汉至南朝时期，随着荆江与汉江三角洲发育，受掀斜构造沉降和科里奥利力的作用，云梦泽主体局限于江汉平原东南，范围随之缩小。

荆江大堤江陵郝穴铁牛矶段
Tieniuji section of the Jingjiang Dike in Haoxue town of Jiangling county

襄阳汉江国家湿地公园　Xiangyang Hanjiang River National Wetland Park

唐宋时期，江汉平原不断淤积抬高，荆江统一河床形成后水位抬升，促使江汉平原湖泊面积逐渐减少；尤其是在宋代时，修建了荆江北侧的堤防，更是迫使荆江向南分流，从而使云梦泽主体逐渐淤积解体，大面积的湖泊不再存在，演变为星罗棋布的小湖泊——江汉湖群。

明清时期是湖北湿地发生重大变化的时期，在自然因素和人为因素的共同作用下，江汉平原经历了历史上最为剧烈的变化。由于荆江和汉江三角洲的不断发育，江汉平原一直在推动江汉湖群向东南推移；清代中后期，太白湖淤塞成为沼泽，而江汉平原中最大的湖泊——洪湖至此形成。

从湿地演替看，湖北湿地演化早期以自然演替为主，湿地自生自灭；春秋战国以后，随着人类对湿地的开发和利用，人类对湿地的影响加大，湿地演替进入自然-人工演替阶段，人工湿地、洲滩湿地逐渐增加，自然湿地不断减少；20世纪50年代以后，人类对两湖平原湿地的改造达到前所未有的地步，湿地演替以人工演替为主，自然湿地遭到大面积围垦；20世纪80年代以后，由于湖泊的围养，人工湿地面积迅速增加，成为湖北面积最大的湿地类型。随着人类活动对湿地的干扰，湿地生态退化态势逐渐显现。

咸宁金桂湖风光　a beautiful sight of Jingui Lake in Xianning City

巫峡风光　a beautiful sight of Wu Gorge

2. 湖北水系与河湖湿地

湖北省水资源丰富，素有"千湖之省"之称。全省地表水资源总量占全国水资源总量的3.5%，位居全国第10位。湖北境内主要河流有长江、汉江，长江自西向东横贯湖北省，汉江流域面积占全省地域面积的33.9%左右。除长江、汉江干流外，省内各级河流河长5千米以上的有4229条，另有中小河流1193条，河流总长5.92万千米，其中河长在100千米以上的河流41条。长江自西向东，流贯省内41个县（市、区），西起巴东县鳊鱼溪河口入境，东至黄梅县滨江出境，流程1061千米。境内的长江支流有汉江、沮水、漳水、清江，其中汉江为长江中游最大支流，在湖北境内由西北趋东南，流经20个县（市、区），由陕西白河县将军河进入湖北郧西县，至武汉汇入长江，流程920千米。

湖北省河流分布表

流域面积	河流数量/条	总长度/千米
≥50平方千米	1232	40000
≥100平方千米	623	28900
≥1000平方千米	61	9200
≥10000平方千米	10	3200

湖泊：全省现纳入湖泊保护名录的湖泊有755个，水面总面积2706.85平方千米。其中，常年水面面积100亩（1亩=666.7平方米）以上湖泊728个，水面总面积2706平方千米。跨省湖泊3个，省内跨市湖泊12个，城中湖103个。

水库：除三峡、丹江口、葛洲坝、陆水和清江隔河岩、高坝洲、水布垭外，全

崇阳青山国家湿地公园　Chongyang Qingshan National Wetland Park

洪湖湿地　Honghu Lake wetland

省建成水库6921座（含电站水库），总库容353亿立方米。其中，大（1）型4座，大（2）型62座，中型285座，小（1）型1217座，小（2）型5353座。大型水库数量居全国首位。

3. 湖北湿地资源概况

湖北省位于长江中游，地处我国南北过渡地带，全省总面积18.59万平方千米。省内地貌类型多样，地理环境复杂，气候条件的水平地带性和垂直地带性差异显著。

全省河流纵横、湖泊密布，水面宽广，素有"千湖之省、鱼米之乡"的美誉，具有十分丰富的湿地资源。湿地分布非常广泛，据第二次全国湿地资源调查结果，全省8公顷以上的湿地总面积为144.50万公顷，占全省总面积的7.8%。其中：河流湿地面积45.04万公顷，占湿地总面积的31.2%；湖泊湿地面积27.69万公顷，占湿地总面积的19.1%；沼泽湿地面积3.69万公顷，占湿地总面积的2.6%；人工湿地面积68.08万公顷，占湿地总面积的47.1%。全省湿地面积占全国湿地面积5360.26万公顷的2.7%，位列全国第11位，中部第1位。

湖北是全球生物多样性较高的地区之一，湿地动植物资源富集，全省湿地植物种类繁多，共有高等植物172科560属1164种，湿地植物优势种主要有芦苇（*Phragmites australis*）、䅟草（*Phalaris arundinacea*）、菰（*Zizania latifolia*）、莲

长江江豚 *Neophocaena asiaeorientalis*

东方白鹳 *Ciconia boyciana*

黑鹳 *Ciconia nigra*

青头潜鸭 *Aythya baeri*

麋鹿 *Elaphurus davidianus*

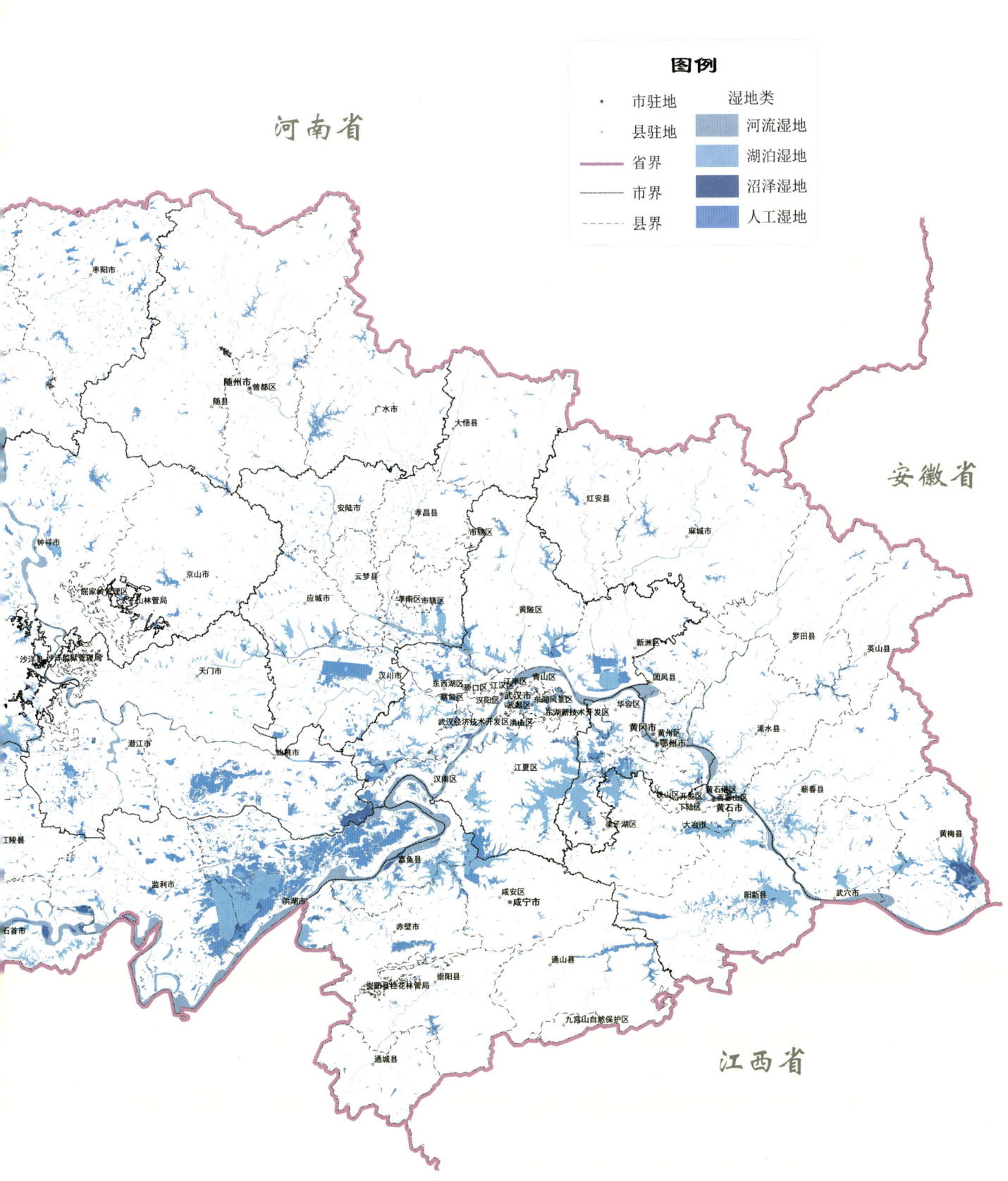

湖北省湿地资源分布图　　distribution of wetland resources in Hubei Province

（*Nelumbo nucifera*）等。湿地区域内分布有国家重点保护野生植物15种，其中一级有水杉（*Metasequoia glyptostroboides*），二级有莼菜（*Brasenia schreberi*）、粗梗水蕨（*Ceratopteris pteridoides*）、水蕨（*Ceratopteris thalictroides*）、野菱（*Trapa incisa*）、莲等14种。

据湖北省第二次湿地资源调查和日常监测结果，结合历史文献资料，湖北省各类湿地生态系统保存野生动物5纲38目108科652种，其中鱼类12目26科201种（亚种），两栖类2目10科68种（亚种）；爬行类2目9科51种；鸟类15目46科285种；哺乳类6目13科47种。属于国家一级重点保护野生动物的有中华鲟（*Acipenser sinensis*）、青头潜鸭（*Aythya baeri*）、长江江豚（*Neophocaena asiaeorientalis*）、麋鹿（*Elaphurus davidianus*）、中国小鲵（*Hynobius chinensis*）等27种；属国家二级重点保护野生动物的有胭脂鱼（*Myxocyprinus asiaticus*）、绢丝丽蚌（*Lamprotula fibrosa*）、巫山巴鲵（*Liua shihi*）、峨眉髭蟾（*Leptobrachium boringii*）、虎纹蛙（*Hoplobatrachus rugulosus*）、乌龟（*Mauremys reevesii*）、水獭（*Lutra lutra*）等81种。

湖北省是东亚雁鸭类迁徙路线、东北亚鹤类迁徙路线和东亚—澳大利亚西亚鸻鹬类迁徙路线的重要组成部分，水鸟资源丰富。有水鸟10目22科162种，其中有45种鸟类被列为国家重点保护动物，属国家一级保护水鸟的有青头潜鸭、中华秋沙鸭（*Mergus squamatus*）、白头硬尾鸭（*Oxyura leucocephala*）、大鸨（*Otis tarda*）、白鹤（*Grus leucogeranus*）、白枕鹤（*Grus vipio*）、白头鹤（*Grus monacha*）、勺嘴鹬（*Eurynorhynchus pygmeus*）、遗鸥（*Larus relictus*）、彩鹳（*Mycteria leucocephalus*）、黑鹳（*Ciconia nigra*）、东方白鹳（*Ciconia boyciana*）、彩鹮（*Plegadis falcinellus*）、黑

池杉　*Taxodium ascendens*

脸琵鹭（*Platalea minor*）、朱鹮（*Nipponia nippon*）、海南鳽（*Gorsachius magnificus*）、黄嘴白鹭（*Egretta eulophotes*）、卷羽鹈鹕（*Pelecanus crispus*）18 种；属国家二级保护水鸟的有鸿雁（*Anser cygnoides*）、白额雁（*Anser albifrons*）、小白额雁（*Anser erythropus*）、红胸黑雁（*Branta ruficollis*）、疣鼻天鹅（*Cygnus olor*）、小天鹅（*Cygnus columbianus*）、大天鹅（*Cygnus cygnus*）、鸳鸯（*Aix galericulata*）、棉凫（*Nettapus coromandelianus*）、花脸鸭（*Sibirionetta formosa*）、斑头秋沙鸭（*Mergellus albellus*）、角䴙䴘（*Podiceps auritus*）、黑颈䴙䴘（*Podiceps nigricollis*）、花田鸡（*Coturnicops exquisitus*）、斑胁田鸡（*Porzana paykullii*）、紫水鸡（*Porphyrio porphyrio*）、蓑羽鹤（*Anthropoides virgo*）、灰鹤（*Grus grus*）、鹮嘴鹬（*Ibidorhyncha struthersii*）、水雉（*Hydrophasianus chirurgus*）、半蹼鹬（*Limnodromus semipalmatus*）、小杓鹬（*Numenius minutus*）、白腰杓鹬（*Numenius arquata*）、大杓鹬（*Numenius madagascariensis*）、翻石鹬（*Arenaria interpres*）、大滨鹬（*Calidris tenuirostris*）、白琵鹭（*Platalea leucorodia*）27 种。湖北省重点保护水鸟 24 种。列入《世界自然保护联盟濒危物种红色名录》（IUCN 红色名录）的有 25 种，列入《濒危野生动植物种国际贸易公约》附录 I 的有 8 种，附录 II 7 种等。

4. 湖北湿地文化资源

自古以来江河湖水滋养了世世代代的湖北人民，秀丽的湿地风光孕育了源远流长的人文历史，丰厚蕴藉的人文历史、优美的传说故事也为湖北湿地增添了更加深邃的内涵和吸引力。全省湿地景观和人文资源丰富，有保存完好的神农架大九湖和咸丰二仙岩高山草甸、泥炭藓沼泽，有著名的三峡风光、清江画廊、三国古战场，有梁子湖、莫愁湖、赤龙湖等美丽的传说。湖北是长江文化的发源地，浸透着"生态长江""文化长江""千湖之省"的独特魅力。

饮食文化。湖北素有"鱼米之乡"之称。境内江河交汇，湖泊遍布，除菱、莲藕外，鱼是湖北最富特色的水产品，青鱼、草鱼、鲢、鳙、鲂、黄颡鱼等经济鱼类在长江流域和江汉湖群广为分布，产量丰富。

水运文化。湖北素有"九省通衢"美誉，境内水系发达，以长江和汉江为骨干，构成了"江汉纳千河"的水系网络和优越的水运条件。

民俗文化。一是渔俗文化，湖区许多村民世世代代以打鱼为生，流传着"赶仗不限地方，打鱼不分廊场"的谚语；二是龙舟文化，端午赛龙舟，即起源于战

恩施大沙坝
Dashaba in Enshi Prefecture

龙舟赛
dragon boat racing

蕲春仙人湖
湿地公园
Qichun Xianren Lake Wetland Park

国时期的秭归县——爱国诗人屈原的故里。

名人与诗词文化。湖北的历史，是人与水共存共荣的历史，浸透着江河湖水魅力的荆楚文化蜚声海内外，以楚三闾大夫屈原为代表，数不尽的文人骚客在此留下无数的诗篇。

湿地风物传说文化。湖北人文底蕴深厚，湿地与周边居民世世代代相生相息，融为一体，还有许多历史事件与湿地息息相关，是对历史文化的折射。

景观文化。湖北湿地自然环境独特。类型丰富、种类多样的湿地景观造就了湖北独特的湿地风光，形成了兼具诗意和灵秀魅力的荆楚风光。

5. 独特的湿地景观

（1）长江及三峡库区。长江由西向东横贯湖北全省，在湖北省全长1061千米。湖北省是长江干线流经里程最长的省份，也是世界上最大的水利工程三峡水利枢纽工程的所在地。三峡工程在正常蓄水位175米淹没线以下形成了中国最大的人工湿地——三峡库区湿地，其中湖北三峡库区湿地总面积16221.81公顷。

（2）汉江及丹江口水库。汉江全长1577千米，为长江第一大支流，其中湖北省境内长920千米。汉江中上游的丹江口水库由1973年建成的丹江口大坝下闸蓄水后形成，2012年大坝加高后水域面积达1022.75平方千米，蓄水量达290.5亿立方米，被誉为"亚洲天池"。丹江口水库是亚洲第一大人工淡水湖，也是国家南水北调中线工程水源地，具有防洪、发电、灌溉、航运、养殖、旅游等综合效益。

丹江口湿地　Danjiangkou wetland

（3）清江。清江为长江一级支流，全长423千米，流域山明水秀，号称八百里清江画廊。清江发源于湖北省利川市齐岳山，流经利川、恩施、宣恩、建始、巴东、长阳、宜都等七个县市，在宜都陆城汇入长江，是长江中游在湖北境内仅次于汉江的第二大支流。

（4）洪湖。洪湖位于湖北省南部洪湖市、监利市之间，长江与东荆河间的洼地中。湖面高程25米，常年水面350平方千米，是湖北省第一大天然淡水湖，2008年被列入国际重要湿地名录，也是国家级自然保护区。

（5）长江天鹅洲故道湿地。长江故道群湿地位于长江中游下荆江河段，江汉平原南缘，荆州石首市与监利市境内，包括天鹅洲、黑瓦屋等7条长江故道，构成了长江流域独特典型的故道群湿地，其中长江天鹅洲故道湿地拥有天鹅洲豚类国家级自然保护区和石首麋鹿国家级自然保护区，维系着长江江豚和麋鹿两个珍稀濒危物种的生存，生态功能与地位突出显著。

（6）大九湖沼泽湿地。大九湖湿地位于湖北省西北端大巴山脉东麓的神农架西南边陲，2013年被列入国际重要湿地名录。大九湖湿地生态系统特殊且典型，其湿地类型包括泥炭藓沼泽、亚高山浅水湖泊、人工库塘、草本沼泽、河流等多种类型，是湖北乃至华中地区目前保存较为完好的亚高山泥炭藓沼泽类湿地，在全国湿地生态系统中具有典型性、特殊性、代表性和稀有性，有极其重要的保护、科研和利用价值。

二、"千湖之省"湖北湿地功能

1. 湖北湿地资源功能

（1）物质生产。湖北省湿地孕育着丰富的动植物资源，可为人类的生产、生活提供大量必需的物质产品。湿地给渔业生产提供了优越的条件，湖北省河流湿地、湖泊湿地、库塘等湿地资源丰富，是全国重要的水产基地；水生经济植物中莲藕产量最高，超过其他经济植物总产量的两倍，莲藕、莲子是湖北省水生经济植物的一大特色，带来了极大的经济效益。此外还有荸荠（*Eleocharis dulcis*）、芡（*Euryale ferox*）、菱（*Trapa bispinosa*）、莼菜（*Brasenia schreberi*）等具有地方特色的水生经济作物。

石首麋鹿国家级自然保护区　Shishou Milu National Nature Reserve

大九湖湿地　Dajiuhu Lake wetland

（2）水源供给、灌溉。长江、汉江及其支流以及大多数湖泊是湖北省居民用水、工业用水、农业用水的水源。湿地是天然的蓄水池，当水从湿地流入蓄水系统时，蓄水层的水就得到了补充，地下水系统为周围地区供水，维持水位或最终流入深层地下水系统，成为长期的水源。

（3）航道运输。湖北有着"九省通衢""千湖之省"的美称，有着得"水"独厚的优势，通航河流229条，通航里程8385千米，居全国第6位，在全国水路交通布局中具有重要的战略地位，是名副其实的水运大省。

（4）能源生产。湖北省江河纵横，水系发达。长江由重庆市巫山县进入湖北省，自西向东横贯全省，至黄梅县出境，湖北省内长1061千米；汉江自陕西白河进入湖北省，至武汉汇入长江，长920千米，沿程有堵河、南河、滚河、唐白河等汇入。湖北得天独厚的区位优势使其蕴藏着十分丰富的水电资源。

举世瞩目的三峡工程、葛洲坝工程、南水北调中线工程水源地丹江口水库、水布垭水电站等，为工农业生产及居民生活提供了充足的电力保障。

2. 湖北湿地生态功能

（1）固碳产氧。湿地生态系统中的绿色植物具有固定大气中的二氧化碳从而减缓地球的温室效应，放出氧气调节大气中的空气组分的功能。湿地固碳包括植被固碳和湿地土壤固碳，已有研究表明湿地土壤固碳能力远远大于湿地植物，湿地土壤单位面积固碳量大约是湿地植物的15倍。

（2）涵养水源，防洪滞沥。水分调节是湿地的重要功能之一。湿地具有巨大的渗透能力和蓄水能力，湿地植物可使降水进入江河的时间滞后，从而达到削洪的目的。湖北省位于长江中游，属亚热带季风性温润气候区，降水主要集中在夏季，每逢暴雨，河水向下游宣泄迅猛异常，易造成水灾。湖北湿地宽阔的水域面积可大大削减洪水的威胁。

（3）滞留沉积物，净化水质。湿地是一个净化器，具有清除杂质和转化毒物的功能。湿地有助于减缓水流的速度，当含有毒物和杂质（农药、生活污水和工业排放物）的流水经过湿地时，流速减慢，有利于毒物和杂质的沉淀和排除。此外，湿地植物能够有效地吸收有毒物质，净化水质。

（4）生物栖息地。水草丛生的湿地环境为鸟类提供了丰富的食物来源，创造了营巢、避敌的良好条件。有一些夏候鸟和冬候鸟把湖北湿地作为其生命循环的一部

枣阳东郊水库
Dongjiao reservoir in Zaoyang city

长江航道
waterway of the Yangtze River

十堰黄龙滩国家湿地公园
Shiyan Huanglongtan National Wetland Park

荇菜　*Nymphoides peltata*

野菱　*Trapa incisa*

野莲　*Nelumbo nucifera*

芡　*Euryale ferox*

睡莲　*Nymphaea tetragona*

分，在迁徙过程中，停歇、休息、取食等要依赖于湿地。水生动物中，长江江豚、中华鲟等要借助湿地繁殖。

3. 湖北湿地人文功能

（1）科教价值。长江、汉江以及丹江口水库、洪湖、梁子湖等湿地具有复杂的湿地生态系统、丰富的动植物群落、珍贵的物种，是科研院所及高等院校用来开展科学研究的重要基地。这些湿地同时也是教育的场所，尤其是进行环境保护、生物多样性、生态系统教育的好场所。

（2）生态旅游价值。目前湖北湿地被作为休闲和旅游的类型，主要是拥有濒危稀有物种、生境、群落、生态系统、景观、自然过程或湿地类型，如长江三峡库区湿地、洪湖湿地、梁子湖湿地、长江天鹅洲豚类国家级自然保护区和石首麋鹿国家级自然保护区等。

三、湖北湿地保护行动与措施

湖北地处长江之"腰"，境内长江、汉江与古云梦泽留下的星罗棋布的湖泊形成了全国最大的江河湖泊淡水湿地复合生态系统，也是全球湿地生物多样性热点地区之一。同时，湖北作为三峡大坝与南水北调中线工程核心水源区所在地，承担着重大且艰巨的生态保护责任，也是长江经济带发展国家战略的重要支点，肩负着长江经济带高质量发展的光荣责任。推进长江大保护、建设美丽长江，既是关系国家发展全局的重大战略，也是确保湖北可持续发展的必然选择。

1. 践行习近平生态文明思想，以新理念引领湿地保护发展

习近平生态文明思想是习近平新时代中国特色社会主义思想的重要组成部分，建设生态文明是中华民族永续发展的长远大计。作为生态文明体制改革的重要探索之一，湿地保护修复是生态文明建设的重要内容，事关国家生态安全，事关经济社会可持续发展，事关中华民族子孙后代的生存福祉。党的十八大以来，习近平总书记多次强调湿地的重要性，对湿地的保护和恢复一直牵挂于心。"共抓大保护、不搞大开发"是习近平总书记关于长江经济带建设系列重要指示和讲话的精神主线与实践要求，也是关于长江经济带建设思想的高度凝练与根本遵循。

湖北省委、省政府坚持"共抓大保护、不搞大开发"和生态优先、绿色发展理

念，积极践行新时期湿地保护和流域生态建设创新思路，将湿地保护与长江大保护、河流湖泊保护等一体部署、一体推进，取得了良好的工作成效。湖北省第十一次党代会提出，实施长江生态环境重大修复工程，有序开展退田还湖还湿，强化三峡、丹江口、清江等库区和洪湖等重点湖泊的保护。湖北省第十二次党代会强调，加强湿地保护，坚决守住流域安全底线，加强长江、汉江、清江流域的上下游统筹、左右岸协同、干支流互动，共抓长江大保护。

湖北省委、省政府坚持全方位保护、全流域修复、全社会参与，部署推进长江大保护"十大标志性战役"、长江经济带绿色发展"十大战略性举措""长江高水平保护十大攻坚提升行动"等，大力实施污染防治等"三大攻坚战"，持续开展碧水保卫战系列行动，实施长江、汉江、清江湖北段的全面禁捕。湖北省委、省政府主要领导亲自抓湿地保护工作。为进一步抓好湿地保护修复，几任省委书记、省长亲自召开洪湖、梁子湖、大九湖湿地保护现场会，协调安排机构建设、资金落实、产权归属等具体事宜，有力地推动了全省湿地保护修复工作。湖北省人大将湖泊、湿地保护立法作为重点事项，湖北省政协将"加强湖北汉江流域江河湖库和湿地生态保护与修复"作为专项民主监督事项，共同推动形成制度、落地见效。

2. 完善湿地保护顶层设计，用最严格最严密的法治保护湿地

（1）做好规划设计。湖北省编制了系列湿地保护的各类规划。根据国家《长江经济带发展规划纲要》《汉江生态经济带发展规划》，编制了《湖北省长江经济带绿色发展"十四五"规划》《湖北省湿地保护修复"十四五"规划》等，将湿地保护纳入全省经济社会发展及相关专项规划进行安排部署。

（2）完善制度设计。湖北省委、省政府颁发了《关于加快推进生态文明体制改革的实施意见》《关于印发湖北省生态保护红线管理办法的通知》《湖北省实施〈党政领导干部生态环境损害责任追究办法（试行）〉细则》《湖北省湿地保护修复制度实施方案》《关于建立健全生态保护补偿机制的实施意见》《湖北省耕地河湖草地休养生息总体方案》，对全省湿地保护修复做出最严格最严密的制度设计，提出了明确的管控目标。覆盖全省的河湖长制、林长制，层层压实各级党委政府的江河湖库和湿地保护责任，强调切实加强河湖湿地保护修复。湖北省质监部门制定了《湖北省重要湿地认定标准》《湖北省健康湿地评价规范》等系列重要文件，规范指导湿地保护修复工作。

（3）健全法规体系。湖北省人大颁布《湖北省湖泊保护条例》《湖北省汉江流域水环境保护条例》《湖北省清江流域水生态环境保护条例》，湖北省政府出台《湖北省湿地公园管理办法》，持续完善江河湖库及湿地保护的省级法规规章，将江河湖库和湿地保护纳入法治化、规范化轨道。

（4）规范湿地征占用管理。严格实行用途管制，占用湿地遵循"先补后占、占补平衡"原则，加大对重大工程占用湿地审核事项的事前、事中和事后监管，确保湿地面积不减少。

（5）建立高规格的湿地管理机构。湖北省已建立国家湿地公园和保护区的管理机构44个，其中正处级5个、副处级30个、科级9个，处级以上机构占81%。目前，全省各级党委政府主抓湿地工作的积极性和主动性空前高涨。

3. 创新机制体制，强化湿地保护管理的责任与绩效考核

（1）湖北省将湿地保护纳入地方政府综合绩效评价内容。把湿地保护、修复和管理情况，纳入自然资源资产负债表编制和领导干部自然资源资产离任审计范围，把湿地面积与湿地保护率作为政府换届的依据，使湿地修复成为重要的政绩导向。对破坏湿地问题突出、保护工作不力、群众反映强烈的地区，湖北省林业局将会同有关部门约谈该地区政府的主要负责人。近年来，湖北省创新机制，落实责任，

湖北天堂湖国家湿地公园　Tiantang Lake National Wetland Park, Hubei

绿色家园　a green home

阳新莲花湖国家湿地公园　　Yangxin Lianhua Lake National Wetland Park

不断强化湿地保护的责任考核，营造良好的湿地保护政绩导向。

（2）湖北省委、省政府把湿地保有量、保护率、保护修复面积作为生态保护的重要内容纳入全省绿色发展指标体系和生态省建设工作考核指标。全省17个市（州）曾有12个市（州）因保护率不达标或国家湿地公园验收未一次性通过而被扣分，引起了各市（州）党委、政府的高度重视，进一步提高了各级党委、政府的湿地保护责任担当。

（3）湿地保护修复纳入了全省林长制、河湖长制考核，以及森林城市、园林城市创建等多项考核的指标体系中。从2015年开始，湖北省委、省政府将湿地管理法规建设与国家湿地公园验收作为市、州、县党政领导班子考核、"三农"考核的指标，强化湿地保护的主体责任与法治意识。

（4）湖北省根据党政领导干部生态环境损害责任追究的相关要求，明确了全省市、县两级党委和政府对本地区生态环境和资源保护负总责，强调"党政同责、一岗双责"，对造成生态严重破坏和资源保护履职不力的责任人实行追责。通过强化问责，倒逼湿地修复责任的落实。

（5）建立湿地资源用途管制机制。2017年湖北省政府办公厅印发的《湿地保护修复制度实施方案》中明确要求对湿地实行负面清单管理，禁止擅自征收、占用国家和省级重要湿地等7种活动，严格限制湿地资源利用强度。凡在年度湿地面积监测中出现面积减少的市州县区，省林业部门将约谈政府主要负责人。

4. 全面实行河湖长制、林长制，推进河湖湿地治理体系和治理能力现代化

湖北是河湖大省，因水得名，因水而兴。湖北在全国率先出台《关于全面推行河湖长制的实施意见》，2017年底全面实行"河湖长制"，不断推动河湖治理体系和治理能力现代化，持续改善河湖面貌和生态环境，建设更多造福人民的幸福河湖。2017年湖北省全面推行河湖长制，湖北省委、省政府主要领导担任省级总河湖长；建立省、市、县、乡、村五级河湖长，同时聘请"民间河湖长"；设立省、市、县三级河湖长制办公室；实现了河湖管护责任全覆盖。湖北省"河湖长制"负责重要河流、湖泊的管理和保护，全省1232条流域面积超过50平方千米的河流和隶属省政府保护的755个湖泊全部纳入其中，完成"一河（湖）一策"的目标。同时，充分考虑湖北河网密布、小微水体众多的现状，将河湖长制责任体系下放至村组一线，将库塘、沟渠这些微小支流也纳入管理范围，力求推动整体生态保护与规范管理。

2021年以来，湖北省贯彻落实中央决策部署，出台《关于全面推行林长制的实施意见》，在全省推行林长制，构建省、市、县、乡、村五级林长组织体系，全

林长公示牌 forest chief public notice board

咸丰县国有坪坝营林场　forest farm of state-owned Pingbaying in Xianfeng county

省1.17亿亩森林每一个山头、每一棵树都有专门的守护者、责任人。秉持"绿水青山就是金山银山"的发展理念，聚焦"护绿、增绿、管绿、用绿、活绿"五大任务，以"林"为重点推深做实林长制。

湖北在推行河湖长、林长制的过程中，强化工作机构、建立部门联动机制、构建长效机制。坚持河长制、湖长制、林长制一体部署、统筹推进，构建了纵向"1 + X"和横向"1+N"模式的组织平台架构。纵向"1 + X"模式，即省、市、县三级河湖长制办公室＋同级工作机构；横向"1+N"模式，即在强化同级河湖长制办公室统筹协调职能的基础上，通过建立同级河湖长制联席会议制度，明确河湖长联系单位职责。出台《关于全面推行林长制的实施意见》，并配套《湖北省林长会议制度》《湖北省林长制信息公开制度》《湖北省林长制部门协作制度》《湖北省林长制督查考核制度》《关于建立"林长＋检察长"协作机制的指导意见》等文件，形成"1+N"制度体系。省级"1+N"制度体系，为林长制改革立柱架梁，形成环环相扣的运行机制。

建立"河湖长、林长＋"工作协作监管机制，充分发挥公安机关、检察机关、审判机关的职能作用，提升河湖与森林资源管理的法治化水平。同时，加强湖北省河湖长、林长制信息化建设，河湖长、林长制信息管理系统建设纳入了湖北省数字政府建设项目，实现河湖森林的科学化、智慧化管理。

5. 强化科技支撑，大力开展湿地保护修复

湿地保护修复涉及整个湿地生态系统与周边社区人文经济发展情况，任务重、难度大、问题多。在全面统筹考虑的基础上，湖北近年来在湿地保护修复工程中强化科技支撑，不断提高湿地治理的生态成效。

（1）依靠科技提高保护修复水平

湖北成立了由43名专家组成的湿地专家委员会，强化湿地修复的技术指导。建立专家服务湿地制度，开展湿地工作定向指导，共享湿地保护管理方面的科研力量，合作开展湿地修复的科研攻关工作。据统计，湖北省已有36处湿地保护区与湿地公园实行了专家合作与联系制度，今后将全覆盖。通过成立专家委员会，建立服务制度，极大地调动了全省湿地专业人士投身湿地保护的积极性与热情，专家走进湿地，科技服务湿地的氛围已经形成。

（2）尊重自然实施退垸还湿

2016年湖北省遭受强降雨洪涝灾害，过半数的湖泊水位超保证，25%的垸堤破溃分洪。全省上下痛定思痛，省委常委会果断决定将鄂州市牛山湖100平方千米的分洪区域作为永久性的湿地保存，实施退垸还湿。同时，湖北省委办公厅、省政府办公厅要求各地将湖泊调蓄的应急措施与退垸还湿地的长远谋划紧密结合起来，大力开展退垸还湿。在面临耕地核减、人员安置难的情况下，全省决定对洪灾中溃口的215个垸堤实施永久性还湖工程。

科学规划退耕还湿。根据国家林业局《退耕还湿实施方案》要求，湖北省林业主管部门编制了《湖北省"十三五"退耕还湿实施方案》，指导全省的退耕还湿工作。积极开展退渔还湿。为追求绿色GDP，全省出重拳取缔围栏、围网、网箱养殖，在天然湖泊湿地和人工库塘湿地积极实行人放天养，强力推行退渔还湿。在长江实行禁渔活动，推进长江湿地生态修复工作。仅在2016年全省江河湖库拆除围栏、围网、网箱面积达5.2万公顷。

（3）完善保护体系，提高湿地保护率

湖北省已连续三年发布省级重要湿地名录，形成了国家重要湿地（国际重要湿地）、省级重要湿地、一般湿地三级管理格局。目前，已建立国际重要湿地4处、数量位居全国第2，国家重要湿地8处、数量位居全国第1，省级重要湿地46处。全省已建立湿地保护区（小区）72个，国家湿地公园66个、数量位居全国第3，

省级湿地公园38个，截至2021年底，湿地保护率达52.6%。

随着一系列湿地保护与修复、退耕还湿、湿地生态效益补偿等项目的有序落实，湖北省湿地生态功能稳步提升，生物多样性持续增加。

6. 积极推进湿地保护国际合作

湖北省加强了与全球环境基金（GEF）、联合国开发计划署（UNDP）、世界自然基金会（WWF）等国际组织合作，在洪湖、涨渡湖、天鹅洲实施了鸟类监测、替代生计等多项湿地生态保护项目；建立了"湖北湿地保护网络"，成功经验被国家林业和草原局推广到长江流域12个省市。

2018年8月，湖北UNDP-GEF湿地项目通过了联合国开发计划署终期评估，国内外独立评估专家对项目实施及成效给予较高评价，项目范围覆盖5个国家级自然保护区和3个省级湿地自然保护区，总投资2081.3万美元，其中GEF赠款265.5万美元、联合国开发计划署配套70.0万美元、湖北省人民政府配套1745.8万美元。项目的实施加速推进了全省湿地保护体系建设，有效保护了湖北省重要湿地的生物多样性。

2019年6月，《湿地公约》常委会审议通过了在武汉举办第十四届缔约方大会的决议，这是在我国首次召开国际性湿地生态保护盛会。本次大会是开展对外交流、系统展示全省生态文明建设成果的重要平台，也是展示湖北省武汉市城市发展新形象的重要窗口。

四、湖北湿地保护管理的重要成效

湖北坚持生态优先、绿色发展，始终把修复长江生态环境摆在压倒性位置，将绿色崛起作为湖北高质量发展的重要底色，践行"两山"理念，为长江大保护交出合格答卷。始终采取强有力措施，持续推进湿地保护修复，取得了良好的生态、社会和经济效益。

1. 湿地生态修复成效明显

"十三五"期间，湖北省累计争取中央财政资金5.15亿元，修复退化湿地面积10.79万亩，实施退耕还湿19.16万亩，新增湿地面积7.77万亩。2021年湖北省启动实施国土绿化五年攻坚提升行动，加快建设长江、汉江、清江生态廊道，全

白琵鹭　*Platalea leucorodia*

府河湿地　Fuhe River wetland

网湖湿地　Wang Lake wetland

年完成营造林196.6万亩。

根据湖北省冬季水鸟调查，冬季水鸟种类从2016年的73种增加到2020年的85种，增加了16%；种群数量从2016年的189212只增加到2020年的651305只，增加了240%。

湖北省河湖水质持续提升。截至2020年，主要河流总体水质为优。179个河流断面中，Ⅰ～Ⅲ类水质断面占93.9%，Ⅳ类断面占6.1%，无Ⅴ类和劣Ⅴ类断面；32个湖库水域中，Ⅰ～Ⅲ类水域占62.5%，Ⅳ类水域占28.1%，Ⅴ类水域占9.4%，无劣Ⅴ类水域。

同时，湖北省开展长江禁捕攻坚战，实现长江干流、汉江干流和水生生物保护区全面禁捕。洪湖国际重要湿地已实施退垸还湿15.88万亩，生态环境明显好转，挺水植物分布面积达10万亩，冬候鸟数量近20万只，生物多样性明显提升。

2. 湿地保护管理评价体系基本建立

（1）湿地保护的分级管理体系基本建立

湖北省制定了《湖北省重要湿地认定标准》，印发了《湖北省湿地名录管理办法》，发布了省级重要湿地名单，形成了国家重要湿地（国际重要湿地）、省级重要湿地、一般湿地三级管理格局。建立了以湿地类型自然保护区（小区）和湿地公园为主体、多种保护形式相结合的湿地保护管理体系。

（2）湿地保护目标责任考核体系基本建立

湖北省政府办公厅出台了《湿地保护修复制度实施方案》，对湖北省湿地保护修复工作进行了总体制度设计，并将湿地面积分解到市县，实行湿地面积总量目标管控。湖北省委省政府把湿地保有量、湿地保护率纳入湖北省绿色发展指标体系和生态省考核目标体系，实行年度考核。全面落实最严格水资源管理制度，水资源管理"三条红线"纳入政府绩效考核体系，对节约用水进行重点监管。

（3）湿地用途监管机制基本建立

2017年湖北省政府办公厅印发的《湿地保护修复制度实施方案》中明确要求对湿地实行负面清单管理，禁止擅自征收、占用国家和省级重要湿地等7种活动，严格限制湿地资源利用强度。规范湿地征占用管理，严格实行用途管制。占用湿地遵循"先补后占、占补平衡"原则，加大对重大工程占用湿地审核事项的事前、事中和事后监管。水域岸线空间管控日趋完善，截至2020年，全省流域面积50平方千米以上的1232条河流、列入省政府保护名录的755个湖泊、6698座水库划界完成，并界定了管理范围。

（4）湿地监测评价体系基本建立

湖北省印发了《湖北省森林湿地资源动态监测体系实施方案》，制定了《湿地资源调查操作细则》，每年开展全省范围的湿地监测工作，每年出具年度湿地监测报告。开展监测监控系统建设。全省已有47个湿地公园建立了视频监控系统，12个国家湿地公园已与湖北省林业局的网络通道建立公文、视频传送的网上传输，部分湿地公园与环保、农业、水利、气象等部门的合作，开展了水质、鱼类、鸟类的专项监测。组织省内国际重要湿地定期开展国际重要湿地生态监测和数据信息更新。每年冬季组织开展全省冬季水鸟同步调查。对青头潜鸭、中华秋沙鸭等标志性物种组织开展专项监测。

3. 开展合理利用湿地特色资源

近年来，湖北省各级湿地管理部门加大湿地资源合理利用的试点示范力度，对各类湿地利用活动进行分类指导，积极开展符合湿地保护要求的生态旅游、生态

灵秀漳河　graceful Zhanghe River

峡江橙香　good harvests of oranges in Xiajiang county

农业、生态教育、自然体验等活动。

神农架大九湖国家湿地公园通过实施退耕还湿、移民搬迁、生态补水等措施修复湿地，在保护好资源的前提下积极发展生态旅游。2018年大九湖国家湿地公园的门票收入3500万元，旅游综合总收入达3.5亿元。

湖北蕲春赤龙湖国家湿地公园积极开展自然教育，2015年被确定为湖北省湿地宣传教育示范基地。2020年创立赤龙湖自然课堂教育品牌，结合民宿主题，打造湿地体验、自然环保、观光休闲于一体的特色生态旅游。建有建筑面积900平方米的自然课堂教室，可以满足40～50人的团体活动需要。同时编撰出版了《赤龙湖国家湿地公园野生动物图鉴》《赤龙湖国家湿地公园生态自然观察手册》《蕲春水生药用植物图鉴及使用指南》等多本科普读物，《赤龙湖国家湿地公园植物彩色图谱》一书获得湖北省科技进步奖三等奖。从2018年下半年开始，每年参加研学旅行的中小学生达8万人次。

武汉东湖风景区　the East Lake Scenic Area in Wuhan

大九湖湿地生态旅游　Dajiu lake wetland ecotourism

湖北省近年来大力建设小微湿地，实现湿地保护与产业经济协同共生。湖北远安徐家庄乡村小微湿地为生态农业型小微湿地，该地发挥自然优势，建设生态农业小微湿地，走"一村一品"特色。通过美丽乡村道路建设，已建成集生态观光、科普教育、教学实习一体的自然教育基地。同时通过虾稻连作，实现农业立体发展，促进农民增收。

4. 社会公众的湿地保护意识增强

湖北省每年以"世界湿地日""爱鸟周"宣传活动为契机，加大湿地保护宣传力度，持续提高全民的湿地保护意识。2022年是《湿地保护法》实施的第一年，湖北省委常委会、省人大常委会专题研究部署《湿地保护法》贯彻实施宣传工作，湖北省人大、省政府召开了贯彻实施《湿地保护法》电视电话会，湖北省人大、省林业局举办了贯彻实施《湿地保护法》新闻通气会，湖北省政府出台贯彻落实《湿地保护法》实施意见。湖北省林业局联合湖北广播电视台、湖北科学技术出版社策划制作出版《大美湿地润荆楚》宣传片和《大美湖北湿地》主题宣传图书（"一片一册"）；组织国家和湖北省主流媒体，分批次采访报道全省湿地生态保护修复成果。湖北省委宣传部、省河湖长制办公室联合印发了《全面推进河湖长制"六进"实施方案》，促进河湖长制走进党校、机关、企业、社区、农村和学校，全社会湿地保护意识明显增强。

秀美香溪河　elegant Xiangxi River

5. 成功申办《湿地公约》第十四届缔约方大会

2019年2月,湖北省武汉市正式启动了承办《湿地公约》第十四届缔约方大会相关工作,通过多方积极努力,在2019年《湿地公约》常委会第57次会议上,审议通过了湖北省武汉市承办《湿地公约》第十四届缔约方大会的议题,2022年6月,经中国政府批准,并经《湿地公约》常委会第59次会议审议通过,决定《湿地公约》第十四届缔约方大会于2022年11月在湖北省武汉市举行。

《湿地公约》第十四次缔约方大会执委会副主任、湖北省林业局党组书记、局长王昌友先后到沉湖和东湖调研《湿地公约》缔约方大会考察现场筹备工作

Wang Changyou, Deputy Director of the Executive Committee of the Convention on Wetlands' COP14, Director of Hubei Forestry Administration has visited Chenhu Lake and East Lake to investigate

Chapter 2

Hubei Wetland Conservation Practices in the Context of Yangtze River Coordinated Conservation

2.1 Environmental evolution and resources of wetlands in Hubei Province

2.1.1 Formation and evolution of wetland in Hubei

Hubei wetland take the Yangtze River and Hanjiang River as the backbone, with interlocking braches, lakes and reservoirs, constituting a complete and diverse wetland ecosystem, maintaining the ecological security and biodiversity of wetlands in the Yangtze River basin and the whole country.

The evolution of wetland ecosystems in Hubei is closely related to the Yangtze River. During the prehistoric period, the Jianghan Plain was a complete wetland ecosystem. From the end of the Late Pleistocene to the beginning of the Holocene, Hubei's wetlands showed a wetland landscape with interlocking rivers and lakes and a small wetland area; the main wetland types were river wetlands and lake wetlands which were mainly formed by natural succession.

During the Spring and Autumn and the Warring States period, Hubei wetlands were mainly distributed in the Jianghan Plain as a plain lake-marsh wetland landscape, with "nine hundred li square" of lakes and marshes. Yunmeng Marsh was formed in the Jianghan plain Lake District. At that time, Yunmeng Marsh was rich in biodiversity, and wetland plants and animals were everywhere. The Jingjiang River had tributaries such as Xiashui and Yongshui that flowed into the Yunmeng Marsh, forming a land delta that extended eastward. In the late Spring and Autumn period, Chu State excavated the abandoned diversion channel to connect with Hanjiang River, which was the artificial canal - Yangshui, while the Lower Jingjiang River was still a lake marsh area at that time. From the Qingjiang and Hanjiang dynasties to the Southern dynasty, as the delta of Jingjiang River and Hanjiang River developed, the scope of Yunmeng Marsh was then reduced since its main body was confined to the southeast of Jianghan Plain by the effect of subsidence of tilting tectonic plate and Coriolis force.

During the Tang and Song dynasties, the Jianghan Plain continuously silted up and the water level was raised after the formation of the unified riverbed of the Jingjiang River, which led to the gradual reduction of the lake area in the Jianghan Plain; especially during the Song dynasty, the dike on the north side of the Jingjiang River was built, forcing the Jingjiang River to divert to the south, thus causing the main body of the Yunmeng Marsh to gradually silt up and disintegrate, and a large area of lakes no longer existed, evolving into a scattering of small lakes - Jianghan lake group.

The Ming and Qing dynasties were a period of major changes for wetlands in Hubei, and the Jianghan Plain experienced the most dramatic changes in its history due to a combination of natural and human factors. As a result of the continuous development of the Jingjiang and Hanjiang River deltas, the Jianghan Plain has been pushing the Jianghan lake group to the southeast; in the middle and late Qing Dynasty, Taibai Lake silted up and became a marsh, while Honghu Lake, the largest lake in the Jianghan Plain, formed here.

From the perspective of wetland evolution, the early stage of wetland evolution in Hubei was mainly natural, and wetlands emerged and perished of itself. After the Spring and Autumn and the Warring States period, with the development and utilization of wetlands by humans, human influence on wetlands increased, and wetland evolution entered the stage of natural-artificial evolution, with artificial wetlands and shoaly wetlands gradually increasing and natural water body wetlands continuously decreasing. After the 1950s, the transformation of the wetlands of the plains reached an unprecedented level, making wetlands succession mainly artificial one and natural wetlands reclaim on a large scale; after the 1980s, due to the enclosure of the lakes, the area of artificial wetlands increased rapidly and became the largest area of wetlands types in Hubei. With the interference of human activities on wetlands, the ecological degradation of wetlands gradually emerged.

2.1.2 Water system and riverine wetlands in Hubei

Known as the "province of a thousand lakes", Hubei Province is rich in water resources. Its total surface water resources account for 3.5% of the total of the country, ranking the 10th in China. The main rivers in Hubei are the Yangtze River and Hanjiang River, with the former crossing Hubei from the west to the east and the area of Hanjiang River basin accounting for about 33.9% of the province's area. In addition to these two rivers, there are 4,229 rivers of more than 5 km in length at all levels in the province, and 1,193 small and medium-sized rivers, with a total length of 59,200 km, including 41 rivers of more than 100 km in length. The Yangtze River flows through 26 counties and cities in the province from the west to the east, entering from the mouth of Bianyu Stream in Badong County in the west to the exit of Binjiang River in Huangmei in the east, with a flow of 1,061 kilometers. The tributaries of the Yangtze River in the territory include Hanjiang River, Jushui, Zhangshui, Qingjiang River, of which Hanjiang River is the largest one of the middle reaches of the Yangtze River, flowing through 20 counties and cities in the territory of Hubei from the northwest to the southeast, that is, from the Jiangjun River in Baihe County, Shaanxi into Yunxi County in Hubei, and to Wuhan into the Yangtze River, with a length of 920 km.

755 lakes in the province are now included in the list of lake conservation, with a total water surface area of 2,706.85km^2. 728 of them possess a perennial water surface area of more than 100 acres with a total water surface area of 2,706km^2. There are 3 inter-provincial lakes, 12 inter-city lakes in the province, and 103 lakes in the city.

In addition to the dams such as Three Gorges, Danjiangkou, Gezhouba, Lushui, Geheyan, Gaozhouba, Shuibuya, the province has completed 6,921 reservoirs (including power station reservoirs), with a total capacity of 35.3 billion m^3. There are 4 large (1), 62 large (2), 285 medium-sized, 1,217 small (1), and 5,353 small (2), making the province the first in number of large reservoirs in the country.

2.1.3 Overview of wetland resources in Hubei

Hubei Province is located in the middle reaches of the Yangtze River, which is also the transition zone between the north and the south of China, with a total land area of 1859×10^4 hm^2. The province has a variety of landform types, a complex geographical environment, and significant zonal differences horizontally and vertically of climate conditions.

Known as the "province of a thousand lakes and the home of fish and rice", the province has a wide range of rivers, lakes, water areas, and rich in wetland resources, The distribution is very extensive. According to the results of the second national wetland resources survey, the total area of wetlands of more than 8 hectares in the province is 1,445,000 hectares, accounting for 7.8% of the provincial land area of 18,590,000 hectares. To be specific, the area of river wetlands is 450,400 hectares, accounting for 31.2% of the total wetland area; the area of lake wetlands is 276,900 hectares, accounting for 19.1% of the total wetland area; the area of marsh wetlands is 36,900 hectares, accounting for 2.6% of the total wetland area; the area of artificial wetlands is 680,800 hectares, accounting for 47.1% of the total wetland area. The wetland area of the province accounts for 2.7% of the national wetland area of 53,602,600 hectares, ranking the eleventh in the country and first in central China.

As a global biodiversity hotspot, Hubei is rich in wetland plant and animal resources. The province enjoys a wide variety of wetland plants, with a total of 1,164 species of higher plants in 172 families and 560 genera, and *Phragmites australis*, *Phalaris arundinacea*, *Zizania latifolia*, and *Nelumbo nucifera* as the dominant species of wetland plants. There are 15 species of national key protected wild plants in the wetland area, including *Metasequoia glyptostroboides*, the species of class I, and *Brasenia schreberi*, *Ceratopteris thalictroides*,

Ceratopteris pteridoides, *Trapa incisa*, *Nelumbo nucifera*, etc, the 13 species of class II.

According to the results of the second wetland resource survey and daily monitoring in Hubei Province, combined with historical literature, the various wetland ecosystems in Hubei Province preserve 652 species of wild animals in 5 classes, 38 Orders and 108 families, including 201 species (subspecies) of fish in 12 orders and 26 families, 68 species (subspecies) of amphibians in 2 orders and 10 families, 51 species of reptiles in 2 orders and 9 families, 285 species of birds in 15 orders and 46 families, and 47 species of mammals in 6 orders and 13 families. The species of mammals are 6 species, 13 families and 47 species. There are 27 species of wildlife under national Class I protection, including *Acipenser sinensis*, *Aythya baeri*, *Neophocaena asiaeorientalis*, *Elaphurus davidianus*, *Hynobius chinensis*, etc. There are 81 species of wildlife under national Class II protection, including *Myxocyprinus asiaticus*, *Lamprotula fibrosa*, *Liua shihi*, *Leptobrachium boringii*, *Hoplobatrachus rugulosus*, *Mauremys reevesii*, *Lutra lutra*, etc.

Hubei Province is an important part of the migration route of East Asian geese and ducks, Northeast Asian cranes and East Asian-Australian plovers, and is rich in water bird resources. There are 162 species of waterfowl in 10 classes and 22 families, among which 45 species have been listed as national key protection animals. There are 18 species of wildlife under national Class I protection, including *Aythya baeri*, *Mergus squamatus*, *Oxyura leucocephala*, *Otis tarda*, *Grus leucogeranus*, *Grus vipio*, *Grus monacha*, *Eurynorhynchus pygmeus*, *Larus relictus*, *Mycteria leucocephalus*, *Ciconia nigra*, *Ciconia boyciana*, *Plegadis falcinellus*, *Platalea minor*, *Nipponia Nippon*, *Gorsachius magnificus*, *Egretta eulophotes*, *Pelecanus crispus*; There are 27 species of wildlife under national Class II protection, including *Anser cygnoides*, *Anser albifrons*, *Anser erythropus*, *Branta ruficollis*, *Cygnus olor*, *Cygnus columbianus*, *Cygnus Cygnus*, *Aix galericulata*, *Nettapus coromandelianus*, *Sibirionetta Formosa*, *Mergellus albellus*, *Podiceps auritus*, *Podiceps nigricollis*, *Coturnicops exquisitus*, *Porzana paykullii*, *Porphyrio porphyria*, *Anthropoides virgo*, *Grus grus*, *Ibidorhyncha struthersii*, *Hydrophasianus chirurgus*, *Limnodromus semipalmatus*, *Numenius minutus*, *Numenius arquata*, *Numenius madagascariensis*, *Arenaria interpres*, *Calidris tenuirostris*, *Platalea leucorodia*. There are 24 species of waterbirds in Hubei Province. There are 25 species listed in the IUCN Red List of Threatened Species (IUCN Red List), 8 species listed in Appendix I and 7 species listed in Appendix II of the Convention on International Trade in Endangered Species of Wild Fauna and Flora.

2.1.4 Cultural resources of wetlands in Hubei

Since ancient times, the rivers and lakes in Hubei have nourished generations of Hubei people, and its beautiful wetland scenery has nurtured a long history of humanities, and its illustrious human history and beautiful legends and stories also have added a deeper connotation and attraction to Hubei wetlands. The province is rich in wetland landscape and humanistic resources, with well-preserved Dajiu Lake in Shennongjia, Erxianyan alpine meadows in Xianfeng, peat moss mire, the famous scenery of Three Gorges, Qingjiang River Gallery, the ancient battlefields of the Three Kingdoms, and beautiful legends of Liangzi Lake, Mochou Lake and Chilong Lake. Hubei is the birthplace of the Yangtze River culture, imbued with the unique charm of "ecological Yangtze River", "cultural Yangtze River" and the "province of a thousand lakes".

Food culture. With the confluence of rivers and lakes, Hubei is known as the "hometown of fish and rice". In addition to *Trapa natans* and lotus roots, fish is the most characteristic aquatic product in Hubei. *Mylopharyngodon piceus*, *Ctenopharyngodon idella*, *Hypophthalmichthys molitrix*, *Hypophthalmichthys nobilis*, *Megalobrama amblycephala*, *Pelteobagrus fulvidraco* and other economic fish are widely distributed and abundantly produced in the Yangtze River basin and the Jianghan lake group.

Water transport culture. Known as "the thoroughfare of nine provinces", the water system in Hubei is well-developed, with the Yangtze River and Hanjiang River as the backbone, forming a "thousands of rivers into Jianghan" water network and superior conditions for water transport.

Folk culture. Firstly, it means fishing culture. Many villagers in the lake area have been making a living by fishing for generations, and the proverb of "Battles are not limited to fields. fishing is not limited to fishing grounds" has been passed down. The second is dragon boat culture. Dragon boat festival originated in Zigui County during the Warring States period - the hometown of the patriotic poet Qu Yuan.

Celebrities and poetry culture. The history of Hubei is the history of the coexistence and co-prosperity of people and water. The Jingchu culture, impregnated with the charm of the rivers and lakes, enjoys reputation at home and abroad; countless literati and poets represented by Qu Yuan, a politician of the Chu State, left countless poems here.

Wetland folklore culture. The humanistic heritage in Hubei is profound, and the wetlands and the surrounding residents have been inseparable for generations. Many historical events are closely related to the wetlands, which reflects the history and culture.

Landscape culture. The natural environment of wetlands in Hubei is unique. The wetland landscape has diverse types, forming a unique wetland scenery in Hubei, and adding poetic and graceful charm to Jingchu scenery.

2.1.5 Unique wetland landscape

2.1.5.1 The Yangtze River and Three Gorges Reservoir Area

The Yangtze River runs across Hubei province from the west to the east, with a total length of 1,062 kilometers. Hubei Province is the one with the longest main line of the Yangtze River and is also the location of the Three Gorges Hydropower-Complex Project, the world's largest hydropower-complex project. The Three Gorges Project has formed China's largest artificial wetlands below the normal water level of 175m flooded line, the Three Gorges Reservoir Wetlands, and the total area of the Three Gorges Reservoir Wetlands in Hubei is 16,221.81 hectares.

2.1.5.2 Hanjiang River and Danjiangkou reservoir

Hanjiang River is the largest tributary of the Yangtze River with the length of 1,577 kilometers, and 920 kilometers of which runs in Hubei Province. The Danjiangkou reservoir in the middle and upper reaches of the Hanjiang River was formed after the Danjiangkou dam, which was built in 1973, was lowered to store water. After the dam was raised in 2012, the water area reached 1,022.75 square kilometers and the water storage volume reached 29.05 billion cubic meters, which is known as the "Asian Heaven Pool". Danjiangkou reservoir is the largest artificial freshwater lake in Asia and the water source of the medium route project of the national South-to-North Water Diversion, with comprehensive benefits of flood control, power generation, irrigation, shipping, breeding and tourism.

2.1.5.3 Qingjiang River

The Qingjiang River is a primary tributary of the Yangtze River, with a total length of 423 kilometers. The basin has lush mountains and clear water, and is known as the 800-li Qingjiang River Gallery. It originates from Qiyue Mountain in Lichuan City, Enshi Prefecture, Hubei Province, and flows through seven counties and cities, including Lichuan, Enshi, Xuanen, Jianshi, Badong, Changyang and Yidu, before merging into the Yangtze River at Lucheng, Yidu. It is the tributary of the middle reaches of the Yangtze River only second to Hanjiang River in Hubei.

2.1.5.4 Honghu Lake

Honghu Lake is located in the south of Hubei Province between Honghu City and Jianli City, a depression between the Yangtze River and the Dongjing River. With a height of 25 meters and a perennial water surface of 350 square kilometers, it is the largest lake in Hubei Province and was listed as a wetland of international importance in 2008, as well as a national nature reserve.

2.1.5.5 Wetlands in Tian-e-zhou Oxbow of the Yangtze River

The wetlands in the oxbow groups of the Yangtze River are located in the middle reaches of the Yangtze River in the lower Jingjiang River, the southern edge of the Jianghan Plain. They are in Shishou City and Jianli County, Jingzhou, including seven oxbows of the Yangtze River, such as Tian-e-zhou and Heiwawu, which constitute unique and typical wetlands of oxbow groups in the Yangtze River basin. The Tian-e-zhou Oxbow of the Yangtze River has the national nature reserves for Tian-e-zhou *Neophocaena asiaeorientalis* and Shishou David's deer, which maintains the survival of the two rare and endangered species of *Neophocaena asiaeorientalis* and *Elaphurus davidianus* with outstanding ecological functions and status.

2.1.5.6 Dajiu Lake peat moss mire

The Dajiu Lake wetlands in Shennongjia is located in the southwestern edge of Shennongjia at the eastern foot of the Daba Mountain Range in the northwestern end of Hubei, and was listed as a wetland of international importance in 2013. The wetland ecosystem of Dajiu Lake is special and typical, with wetland types such as peat moss mire, subalpine shallow lakes, artificial reservoirs and ponds, marshes and rivers. It is a relatively well-preserved subalpine peat moss mire-like wetland in Hubei and even in central China, which is typical, special, representative and rare in the national wetland ecosystem and has extremely important values in conservation, scientific research and utilization.

2.2 Functions of wetlands in Hubei—"province of a thousand lakes"

Rich water resources; diverse wetland products; water conservation and flood control and storage; biological habitat and diversity; ecological tourism and environmental education.

2.2.1 Resource function of wetlands in Hubei

2.2.1.1 Material production

The wetlands in Hubei are rich in plant and animal resources, which can provide a large number of essential material products for human production and life. Wetlands provide superior conditions for fishery production. Hubei Province is rich in river wetlands, lake wetlands and reservoirs and ponds, and is an important aquaculture base in the country. Among aquatic economic plants, lotus root has the highest production, more than twice the total production of others. Lotus root and lotus seed are a major feature of aquatic economic plants in Hubei Province, bringing great economic benefits. In addition, there are other aquatic economic crops with local characteristics such as *Eleocharis dulcis*, *Euryale ferox*, *Trapa bispinosa*, *Brasenia schreberi*.

2.2.1.2 Water supply and irrigation

The Yangtze River, Jianghan River and its tributaries, as well as most lakes, are the water sources for residents, industry and agriculture in Hubei Province. Wetlands are natural reservoirs. When water flows from wetlands to the aquifer system, the water in the aquifer is replenished and the groundwater system supplies water to the surrounding areas, maintaining the water level or eventually flowing into the deep underground system level, becoming a long-term water source.

2.2.1.3 Waterway transportation

Known as the "thoroughfare of nine provinces" and the "province of a thousand lakes", Hubei has a unique "water" advantage, namely, 229 navigable rivers with navigable mileage of 8,385 km, ranking the sixth in the country. It has an important strategic position in the national waterway transportation layout, making it a veritable water transport province.

2.2.1.4 Energy production

Hubei Province is crisscrossed by rivers and has a well-developed water system. The Yangtze River enters the province from the Wushan Mountain, Sichuan, and runs across the province from the west to the east to Huangmei County, totaling 1,061km. The Hanjiang River enters Hubei Province from the Baihe River, Shaanxi, and merges into the Yangtze River at Wuhan, totaling 920km, with the confluence of the Duhe River, the Nanhe River, the Gunhe River and the Tangbai River. The unique location advantage makes Hubei rich in hydropower resources.

The Three Gorges Project, Gezhouba Project, Danjiangkou reservoir (the water source of the medium route project of the national South-to-North Water Diversion), and Shuibuya Power Station provide sufficient electricity for industrial and agricultural production and people's life.

2.2.2 Ecological function of wetlands in Hubei

2.2.2.1 Carbon sequestration and oxygen production

The green plants in the wetland ecosystem have the function of fixing the CO_2 in the atmosphere to relieve the greenhouse effect of the earth and release oxygen to regulate the air components in the atmosphere. At the same time, wetland soils have a strong effect of carbon sequestration, and studies have shown that the carbon sequestration capacity of wetland soils is much greater than that of wetland plants. The carbon sequestration quantity of wetland soils per unit is about 15 times more than that of wetland plants. Therefore, wetland carbon sequestration includes vegetation carbon sequestration and wetland soil carbon sequestration.

2.2.2.2 Water conservation, floodwater control

Moisture regulation is one of the important functions of wetlands. It also has huge infiltration capacity and water storage capacity, and wetland plants can lag the time of precipitation entering rivers, so as to achieve the purpose of flood reduction. Hubei Province is located in the middle reaches of the Yangtze River, falling into a subtropical monsoon humid climate zone. Thus the precipitation is mainly in the summer, and the river water washes downstream violently whenever there is a rainstorm, resulting easily in floods. The wide water area of wetlands in Hubei can greatly reduce the threat of downstream flooding.

2.2.2.3 Detention of sediments, purification of water quality

Wetlands is a purifier, with the function of removing and transforming toxic substances and impurities. Wetlands help slow down the speed of water flow, and when the flow of water containing toxins and impurities (pesticides, domestic sewage and industrial emissions) runs through the wetlands, it will be easy to precipitate and eliminate toxins and impurities if it slows down. In addition, some wetland plants can effectively absorb toxic substances.

2.2.2.4 Biological habitat

The wetland environment with water and grasses provides birds with rich food sources and favorable conditions for nesting and avoiding enemies. Some summer and winter migratory birds use the wetlands in Hubei as part of their life cycle and depend on them for resting and feeding during their migration. Among aquatic animals, *Neophocaena asiaeorientalis* and *Acipenser sinensis* breeding in wetlands.

2.2.3 Humanistic function of wetlands in Hubei

2.2.3.1 Scientific and educational value

The Yangtze River, Jianghan and lakes are important scientific research sites for the study of wetland organisms. The Yangtze River, Hanjiang River and Danjiangkou reservoir, Honghu Lake, Liangzi Lake and other wetlands are important bases for scientific research institutes and colleges and universities to conduct scientific research. These wetlands are also perfect for education, especially for environmental conservation, biodiversity and ecosystem education.

2.2.3.2 Value of ecotourism

The types of wetlands in Hubei that are currently used for recreation and tourism are mainly those with endangered and rare species, habitats, communities, ecosystems, landscapes, natural processes or wetland types, such as the wetlands in Yangtze River Three

Gorges Reservoir, Honghu Lake, wetland in Liangzi Lake, national nature reserves for Tian-e-zhou *Neophocaena asiaeorientalis* and for Shishou *Elaphurus davidianus*.

2.3 Conservation action and measures of wetlands in Hubei

Hubei is located at the "waist" of the Yangtze River. And the Yangtze River, Hanjiang River and ancient Yunmeng Marsh leave a scattering of lakes, forming the largest river, lake and freshwater wetland complex ecosystem in the country, and a global biodiversity hotspot of wetlands. At the same time, Hubei, as the location of the Three Gorges Dam and the core water source area of the medium route project of the South-to-North Water Diversion, bears a significant and arduous responsibility for ecological conservation, and a glorious responsibility for the quality development of the Yangtze River Economic Belt as an important pivot point of the national strategy for the development of the Yangtze River Economic Belt. Promoting the well-coordinated environmental conservation of the Yangtze River and building a beautiful Yangtze River is not only a major strategy related to the overall development of the country, but also an inevitable choice to ensure the sustainable development of Hubei.

2.3.1 Practice Xi Jinping's thought of ecological civilization and lead the development of wetland conservation with new concepts

Xi Jinping's thought of ecological civilization is an important part of Xi Jinping's thought of socialism with Chinese characteristics in the new era, and building ecological civilization is a millennium plan for the sustainable development of the Chinese nation. As one of the important explorations of the reform of ecological civilization system, wetland conservation and restoration is an important part of the construction of ecological civilization, a matter of national ecological security, sustainable economic and social development, and the survival and well-being of future generations of the Chinese nation. Since the 18th National Congress of CPC, General Secretary Xi Jinping has repeatedly stressed the importance of wetlands, and deeply concerned about conservation and restoration of wetlands. "Promoting well-coordinated environmental conservation and avoiding excessive development" is the gist and practical requirements of the series of important instructions and speeches of General Secretary Xi Jinping on the construction of the Yangtze River Economic Belt, which is also a highly condensed expression and complete obedience of the ideas on the construction of the Yangtze River Economic Belt.

The provincial Party Committee and Provincial Government of Hubei adhere to the "promoting well-coordinated environmental conservation and avoiding excessive development" and the concept of ecological priority and green development, and actively practice the innovative ideas of wetland conservation and construction of basin ecology in the new era, deploying and advancing wetland conservation, well-coordinated environmental conservation of Yangtze River, river and lake conservation in an integrated manner, which has produced good results. The 11th Party Congress of the province proposed to implement a major restoration project of the ecological environment of the Yangtze River, orderly return land to lakes and wetlands, and strengthen the conservation of the Three Gorges, Danjiangkou, Qingjiang River and other reservoir areas and key lakes such as Honghu Lake. The 12th Party Congress of the province stressed the importance of strengthening wetland conservation, and we should resolutely guard the bottom line of basin safety, strengthen the upstream and downstream coordination of the Yangtze River, Hanjiang River and Qingjiang River basins, synergy between the left and right bank, interaction between main and tributary of the Yangtze River basin, and jointly enhance the well-coordinated environmental conservation of the Yangtze River.

The Provincial Party Committee and the Provincial Government insist on all-round conservation, whole-basin restoration, and whole-society participation, make deployment to promote the "ten landmark battles" of the well-coordinated environmental conservation of the Yangtze River, "ten strategic initiatives" of green development of the Yangtze River Economic Belt, and "ten major elevation initiatives of high-level conservation of the Yangtze River", fight the "three major battles" to vigorously implement pollution prevention and control, continue to implement a series of actions to defend the blue water, and a comprehensive ban on fishing of the northern sections of the Yangtze River, Hanjiang River and Qingjiang River. The main leaders of the Provincial Party Committee and Provincial Government focus on the work of wetlands. In order to further improve the wetland conservation and restoration, several provincial Party committee secretaries and governors personally held on-site meetings on the wetland conservation of Honghu Lake, Liangzi Lake and Dajiu Lake, coordinating arrangements for institutional construction, funding, property rights ownership and other specific matters, which strongly promoted the wetland conservation and restoration work of the province. The Provincial People's Congress takes legislation of lake and wetland conservation as a key matter; the Provincial CPPCC takes "strengthening the ecological conservation and restoration of rivers, lakes, reservoirs and wetlands in the Hanjiang River basin in Hubei" as a special matter for democratic supervision, jointly promoting the formation of the system and effective implementation.

2.3.2 Improve the top-level design of wetland conservation and protect wetlands with the strictest and most stringent rule of law

Make good planning and design. The province has prepared a series of planning for wetland conservation. According to the national Outline of Yangtze River Economic Belt

Development Plan and Hanjiang River Ecological and Economic Belt Development Plan, the 14th Five-Year Plan for Green Development of the Yangtze River Economic Belt in Hubei Province and 14th Five-Year Plan for Wetland Conservation and Restoration in Hubei Province have been prepared, and wetland conservation has been incorporated into the economic and social development and related projects of the province for arrangement and deployment.

Improve design of the system. The Provincial Party Committee and the Provincial Government issued the *Implementation Opinions on Accelerating the Reform of Ecological Civilization System*, *Notice on Issuing Management Measures of Ecological Conservation Redlines in Hubei Province*, *Rules for Implementing Measures for Pursuing the Responsibility of Party and Government Leaders for Damage to the Ecological Environment* (*for Trial Implementation*), *Implementation Plan of Wetland Conservation and Restoration System in Hubei Province*, *Implementation Opinions on Establishing a Sound Compensation Mechanism of Ecological Conservation*, *Overall Plan for Regeneration of Croplands, Rivers, Lakes and Grasslands in Hubei Province*, making the strictest and most stringent institutional design for the conservation and restoration of wetlands in the province, putting forward clear control objectives, covering the river and lake chief scheme and forest chief scheme, so that Party committees and governments at all levels assume definite responsibility for the conservation of rivers, lakes and reservoirs and wetlands, and emphasize the effective strengthening of conservation and restoration of rivers, lakes and wetlands. The provincial quality control department has developed a series of important documents such as Identification Standards for Important Wetlands in Hubei and Evaluation Specifications for Healthy Wetlands in Hubei to standardize and guide wetland conservation and restoration.

Perfect regulatory system. The Provincial People's Congress promulgated the *Regulations on Lake Conservation in Hubei Province*, *Regulations on Water Environmental Conservation of Hanjiang River Basin in Hubei Province*, *Regulations on Water Ecological Environment Conservation of Qingjiang River Basin in Hubei Province*, and the Provincial Government issued the *Management Measures of Wetland Parks in Hubei Province* to continuously improve the provincial regulations and rules on the conservation of rivers, lakes, reservoirs, and wetlands, bringing the conservation of rivers, lakes, reservoirs, and wetlands into the rule of law and standardization.

Standardize management of wetland acquisition and occupation. The use of wetlands should be strictly controlled, and the occupation of wetlands should be in line with the principle of "filling before occupying, filling and occupying in balance"; the supervision of audit before, during and after the occupation of wetlands in major project should be enhanced to ensure that the wetland area is not reduced.

Establish high-profile wetland management institutions. The province has established 44 management agencies of national wetland parks and protected areas, including 5 at the division level, 30 at the deputy division level, 9 at the section level, 81% of the agencies above the division level. At present, the enthusiasm and initiative of Party committees and governments at all levels in the province to take charge of the work of wetlands has never been higher. The leaders at all levels of the Provincial Government grasp, wetlands restoration pattern has been formed.

2.3.3 Innovate mechanism and system and increase accountability and performance assessment of wetland conservation and management

Hubei incorporates wetland conservation into the comprehensive performance assessment of local government. Wetland conservation, restoration and management is included into the balance sheet compilation of natural resources and audit scope of natural resources assets when leaders and cadres leave office, and wetland area and wetland conservation rate are considered one basis for the change of government, so that wetland restoration has become an important political performance. Once there is a region with prominent problems such as wetland destruction, ineffective conservation work, or public complaint, the provincial forestry administration will interview the main responsible person with the relevant departments. In recent years, Hubei has been innovating mechanisms to make leaders be fully accountable, and constantly strengthening performance assessment of wetland conservation in order to make wetland conservation performance-oriented.

The Provincial Party Committee and Provincial Government have incorporated wetland retention, conservation rate, and conservation and restoration area, an important element of ecological conservation, into the green development index system and work assessment indicators for construction of ecological province. There was deduction of points for 12 cities (prefectures) out of 17 because they did not meet the standard of conservation rate or their failed to pass national wetland parks acceptance in one go, causing the high attention of Party Committees and governments of all municipalities (prefectures), further increasing the accountability of the Party Committees and governments at all levels of wetland conservation.

Wetland conservation and restoration has been incorporated into the assessment of forest chief scheme, river and lake chief scheme, and the index system of building forest cities and garden cities. Since 2015, the Provincial Party Committee and Provincial Government have been making construction of wetland management regulations and national wetland park acceptance as the index to assess the Party and government leadership team at city, prefecture and county level and their work on agriculture, rural areas and wellbeing of farmers to strengthen the main responsibility of wetland conservation and awareness of the rule of law.

According to the requirements of *Pursuing the Responsibility of Party and Government Leaders for Damage to the Ecological Environment*, Hubei Province makes it clear that the Party Committees and governments at municipal and county-levels are responsible for the conservation of the ecological environment and resources in the region, emphasizing the "Party and government together, double responsibilities for one job" and pursuing responsibility of the person who causes serious ecological damage or performs ineffective resource conservation. We should force the implementation of wetland restoration responsibility by strengthening accountability.

Establish control mechanism for use of wetland resources. In 2017, the Provincial Government office issued the *Implementation Plan of Wetland Conservation and Restoration System*, clearly requiring the implementation of negative list management of wetlands, prohibiting seven kinds of activities such as unauthorized expropriation, occupation of wetlands of national and provincial importance, strictly limiting the intensity of use of wetland resources. Whenever there appears reduction of wetland area in a city, prefecture, county or district during annual monitoring of wetland area, the provincial forestry department will interview the main responsible person of the government.

2.3.4 Fully implement the river and lake chief scheme and promote the modernization of the riverine wetlands governance system and governance capacity

As a large province of rivers and lakes, Hubei is named because of water, and prosperious because of water. It is the first in the country to introduce Opinions on Implementation of River and Lake Chief Scheme, and fully implement the scheme at the end of 2017, constantly promoting the modernization of river and lake governance system and governance capacity, improving the appearance of rivers and lakes and ecological environment, builing more happy rivers and lakes for the benefit of the people. In 2017, the river and lake chief scheme was fully introduced in Hubei province with the main leaders of the Provincial Party Committee and the Provincial Government as the provincial chief of rivers and lakes; five levels of river and lake chiefs at the provincial, municipal, county, township, and village levels have been confirmed while civil river and lake chiefs have been employed; river and lake chief scheme offices at the provincial, municipal and county levels have been established, to achieve full coverage of river and lake management and care responsibilities. The River and Lake Chief Scheme in Hubei Province is responsible for the management and conservation of important rivers and lakes, which include 1,232 rivers with an area of more than 50 square kilometers and 755 lakes under the conservation of the Provincial Government, to complete the goal of "one river (lake) one policy". At the same time, Hubei takes into full consideration of its current situation of dense river network and numerous small and micro water bodies, and decentralizes the responsibility system of the river and lake chief scheme to the front line of villages and groups, incorporating tiny tributaries of reservoirs and ponds and ditches in the management in an effort to promote overall ecological conservation and standardized management.

In 2021, Hubei Province introduced the Opinions on Full Implementation of Forest Chief Scheme, following the decisions and deployment of the central government. It carries out the forest chief scheme throughout the province, builds five levels of organization system of forest chiefs at the provincial, municipal, county, township, and village levels, making sure every hill and every tree has a special guardian, the person responsible, in the 117 million mu of forest of the province. Hubei upholds the development concept of "lucid water and lush mountains are invaluable assets", focusing on five major tasks of "green conservation, green increase, green management, green utilization, and green vitality" to push forward the implementation of forest chief scheme with the emphasis on forest.

In the implementation of the river and lake chief scheme and forest chief scheme, Hubei strengthens the work of institutions, establishes departmental linkage mechanism, and builds a long-term mechanism. It adheres to the river chief scheme, lake chief scheme, and forest chief scheme with integrated deployment and promotion, and builds an organizational platform structure of vertical "1 + X" and horizontal "1 + N" model. The vertical "1 + X" model means the three levels of river and lake chief scheme offices at the provincial, municipal and county levels plus the work agencies at the same level, while the horizontal "1 + N" model, establishment of the river and lake chief scheme joint meeting system at the same level based on strengthening coordination function of river and lake chief scheme offices at the same level to clarify the responsibilities of the river and lake chief units. The introduction of *Opinions on Full Implementation of Forest Chief Scheme* is complemented by the supporting documents of "*Conference System of Forest Chief in Hubei*", "*Information Disclosure System of Forest Chief Scheme in Hubei*", "*Departmental Collaboration System of Forest Chief Scheme in Hubei*", "*Supervision and Assessment System of Forest Chief Scheme in Hubei*", "*Guiding Opinions on the Establishment of Collaboration Mechanism of 'Forest Chief + Chief Procurator'*", forming the "1+N" institutional system. The "1+N" institutional system at the provincial level erects pillars and beams for the reform of forest chief scheme, forming interlocking operational mechanisms.

The collaborative supervision mechanism of "river and lake chief, forest chief +" should be established to give full play to the role of public security organs, procuratorial organs, and trial organs, to enhance the rule of law of river and lake and forest resources management. At the same time, informatizaion construction of river and lake chief scheme and forest chief scheme should be strengthened; construction of information management systems of river and lake chief scheme and forest chief scheme should be included into the construction projects of digital government of Hubei Province to achieve scientific and intelligent management of rivers, lakes and forests.

2.3.5 Strengthen scientific and technological support and vigorously carry out wetland conservation and restoration

Wetland conservation and restoration involves the entire wetland ecosystem and the humanistic and economic development of surrounding community, making the task heavy, difficult and problematic. On the basis of comprehensive and integrated consideration, Hubei has been strengthening scientific and technological support in wetland conservation and restoration projects in recent years and constantly improving the ecological effectiveness of wetland management.

Improve the level of conservation and restoration with science and technology. Hubei set up a wetland expert committee of 43 experts to strengthen the technical guidance of wetland restoration. It establishes the system of experts serving the wetlands, carries out orientation guidance for wetland work, and cooperation and sharing of scientific research power in wetland conservation and management, and actively engages in scientific research work on wetland restoration. According to statistics, 36 wetland reserves and wetland parks in the province have implemented the expert cooperation and contact system, and the others will also be covered. Through the establishment of expert committees and service systems, the motivation and enthusiasm of wetland professionals has greatly been mobilized to participate in wetland conservation. The pattern of experts into wetlands and serving wetlands with science and technology has been formed.

Respect nature by returning land for levee to wetlands. The year of 2016 suffered heavy rainfall and flooding, making more than half of the lake level exceed the guarantee and 25% of the levees breach. The province learned a lesson from the pain. The standing committee of the Provincial Party Committee decided to preserve the 100-square-kilometer flood diversion zone in Niushan Lake, E-zhou City as a permanent wetland, and return the land for levee to wetlands. At the same time, the General Office of the Provincial Party Committee and the General Office of the Provincial Government asked governments at all levels to closely combine the emergency measures of lake storage with the long-term planning of returning the levees to wetlands to vigorously carry out the long-term planning. Confronted with the reduction of arable land and difficulties in personnel resettlement, the province decided to implement the project of permanent returning to lakes as for the 215 levees breached in flooding. Scientific planning should be made to return land for farming to wetlands. According to the requirements of *Implementation Plan of Returning Land for Farming to Wetlands* by National Forestry Administration, the provincial forestry authorities prepared the *Thirteen Five-Year Implementation Plan of Returning Land for Farming to Wetlands in Hubei Province*, guiding the work of returning land for farming to wetlands in the province. Returning land for fishing to wetlands should also be actively carried out. In order to pursue green GDP, the province makes great effort in outlawing fences, seine nets and cage farming, and actively implements the practice of natural cultivation with human assistance in natural lakes and wetlands in artificial reservoirs and ponds in order to strongly promote the returning of land for fishing to wetlands. The fishing activities are banned in the Yangtze River to promote the ecological restoration of the wetlands in the Yangtze River. In 2016 alone, the province dismantled 52,000 hectares of fences, seine nets and cages of rivers, lakes and reservoirs.

Improve conservation system and increase the rate of wetland conservation. Hubei has released its list of wetlands of provincial importance for three consecutive years, forming a three-tier management pattern of wetlands of national importance (international importance), provincial importance and general importance. At present, four wetlands of international importance have been established, ranking the 2nd in the country in terms of number; eight wetlands of national importance have been established, ranking the 1st in the country in terms of number; and 46 wetlands of provincial importance have been established. The province has established 72 wetland reserves (plots), 66 national wetland parks, ranking the 3rd in the country in terms of number, and 38 provincial wetland parks. By the end of 2021, the rate of wetland conservation reached 52.6%.

Improve survey and monitoring level of wetland resources. An intelligent perception system should be set up to carry out wetland resource survey and monitoring management. Hubei Province has formulated and issued the *Implementation Plan of Dynamic Monitoring System of Forest and Wetland Resources in Hubei Province*, formulated the *Operation Rules on Survey of Wetland Resources*, carried out province-wide wetland monitoring work every year and issued annual wetland monitoring reports every year. Forty-seven wetland parks in the province have established video monitoring and intelligent management systems, and some wetland parks have carried out special monitoring and data sharing of water quality, fish and birds in cooperation with environmental conservation, agriculture, water conservancy and meteorology departments. They carry out identification of species through sound, remote tracking and bird ringing for key species in key areas, and organize regular information update of ecological monitoring and data of wetlands of international importance in the province, synchronized surveys of waterbirds in the province in winter every year, and high-tech special monitoring of iconic species such as *Aythya baeri* and *Mergus squamatus*.

With the orderly implementation of a series of projects for wetland conservation and restoration, return of land for farming to wetlands, compensation of wetland ecological benefits, the ecological functions of wetlands in Hubei Province have been steadily improved and biodiversity has continued to increase.

2.3.6 Actively promote international cooperation in wetland conservation

Hubei Province has strengthened cooperation with international organizations such as Global Environment Facility (GEF), the United Nations Development Programme (UNDP), and the World Wide Fund for Nature (WWF), investing more than 45 million yuan to implement a number of wetland ecological conservation projects such as bird monitoring, alternative livelihoods in Honghu Lake, Zhangdu Lake, and Tian-e-zhou and establish the "conservation network of wetlands in Hubei". The successful experience has been promoted by the National Forestry Administration to 12 provinces and cities in the Yangtze River basin.

In August 2018, the Hubei UNDP-GEF wetland project passed the final evaluation of the United Nations Development Programme, and the independent evaluation experts at home and abroad gave high praise to its implementation and effectiveness. The project covers five national nature reserves and three provincial wetland nature reserves, with a total investment of USD 20.813 million including USD 2.655 million from the GEF grant, supportive USD 700,000 from the UNDP, and USD 17.458 million from Hubei Provincial Government. The implementation of the project has accelerated the construction of the wetland conservation system and effectively protected the biodiversity of important wetlands in Hubei Province.

In June 2019, the Standing Committee of the *Ramsar Convention* considered and adopted the resolution to hold the 14th Conference of the Contracting Parties in Wuhan, which is the first time to hold this international event of wetland ecological conservation in China. The conference is an important platform to carry out foreign exchanges and systematically display the achievements of construction of ecological civilization in the province, as well as an important window to show the new image of Wuhan, Hubei Province, a city revitalized after the epidemic.

2.4 Important results of wetland conservation and management in Hubei

Hubei adheres to the ecological priority and green development, always putting restoration of the ecological environment of the Yangtze River in the overriding position, the rise of green as an important color of quality development in Hubei. It practices the concept of "two mountains", and delivers a qualified answer sheet for the well-coordinated environmental conservation of the Yangtze River. By taking strong measures to continue to promote wetland conservation and restoration, Hubei has achieved much ecological, social and economic benefit.

2.4.1 Clear results of wetland ecological restoration

During the 13th Five-Year Plan period, Hubei province strived for a total of 515 million yuan of central financial funds to restore degraded wetlands with an area of 107,900 mu, return 191,600 mu of farmland to wetlands, and add 77,700 mu of wetlands. In 2021, Hubei province launched the implementation of a five-year campaign to improve the greening of the country, accelerating the construction of the ecological corridors of the Yangtze River, Hanjiang River and Qingjiang River, and completing 1.966 million mu of afforestation throughout the year.

According to the winter waterbird survey of the province, winter waterbird species increased from 73 in 2016 to 85 in 2020, an increase of 16%; the population size increased from 189,212 in 2016 to 651,305 in 2020, an increase of 3.4 times. The water quality of rivers and lakes in Hubei province continues to improve.

By 2020, the overall water quality of major rivers is excellent. Among 179 river sections, I to III water quality sections and IV accounted for 93.9% and 6.1%, and there were no V and inferior V sections; among 32 lake and reservoir waters, I to III waters and IV accounted for 62.5% and 28.1%, V for 9.4%, and there were no inferior V waters.

At the same time, Hubei Province launched the "battle" against fishing in the Yangtze River to achieve a comprehensive ban on fishing in the main streams of the Yangtze River and Hanjiang River and aquatic life reserves. The wetlands of international importance of Honghu Lake has returned 158,800 mu of levees to wetlands, significantly improving the ecological environment. The distribution area of water-holding plants has reached 100,000 mu; the number of winter migratory birds has been nearly 200,000; the biodiversity has improved significantly.

2.4.2 Basic establishment of evaluation system of wetland conservation management

The graded management system of wetland conservation has been basically established. Hubei formulated Identification Standards for Important Wetlands in Hubei, issued "Management Measures of Wetland List of Hubei" and the list of weetlands of provincial importance, forming a three-tier management pattern of wetlands of national importance (international importance), provincial importance and general importance. At present, four wetlands of international importance have been established, ranking the 2nd in the country in terms of number; eight wetlands of national importance have been established, ranking the 1st in the country in terms of number; and 46 wetlands of provincial importance have been established. The province has established a wetland conservation and management system with wetland type as nature reserve (plot) and wetland park as the main body with combination of various forms of conversation, and 72 wetland reserves (plots), 66 national wetland parks, ranking the 3rd in the country in terms of number, and 38 provincial wetland parks. By the end of 2020, the rate of wetland conservation reached 52.6%.

Responsibility assessment system of wetland conservation target has been basically established. The General Office of the Provincial Government issued the Wetland Conservation and Restoration System Scheme, making an overall system design of wetland conservation and restoration work, and dividing wetland areas between cities and counties, so that the total wetland area target could be controlled. The Provincial Party Committee and Provincial Government include wetland retention and rate of wetland conservation into the green development index system and assessment system for construction of ecological province and carry out annual assessment. The province comprehensively implements the strictest management system of water resources, and includes "three redlines" of water resources management into the assessment system of government performance, paying high attention to the supervision of water conservation.

Regulatory mechanism of wetland use has been basically established. In 2017 the General Office of Provincial Government issued the Wetland Conservation and Restoration System Scheme, clearly requiring the implementation of negative list management of wetlands, prohibiting seven kinds of activities such as unauthorized expropriation, occupation of wetlands of national and provincial importance, strictly limiting the intensity of use of wetland resources. The occupation of wetlands should be in line with the principle of "filling before occupying, filling and occupying in balance"; the supervision of audit before, during and after the occupation of wetlands in major projects should be enhanced. Spatial control of water shorelines is increasingly improved. As of 2020, demarcation of 1,232 rivers with basin area of more than 50 square kilometers in the province, 755 lakes included in the conservation list of Provincial Government, and 6,698 reservoirs was completed, and the scope of management was defined.

Monitoring and evaluation system of wetlands has been basically established. The *Implementation Plan of Dynamic Monitoring System of Forest* and *Wetland Resources in Hubei Province* has been issued; and the *Operation Rules on Survey of Wetland Resources* has been formulated; the province-wide wetland monitoring work is carried out every year, and the annual wetland monitoring report is issued every year. The construction of monitoring systems is carried out. Forty-seven wetland parks in the province have established video monitoring systems; 12 national wetland parks have established online transmission of official documents and video transmission with the network channel of the provincial forestry administration and some wetland parks have carried out special monitoring of water quality, fish and birds in cooperation with environmental conservation, agriculture, water conservancy and meteorology departments. They organize regular information update of ecological monitoring and data of wetlands of international importance in the province, synchronized surveys of waterbirds in the province in winter every year, and special monitoring of iconic species such as *Aythya baeri* and *Mergus squamatus*.

2.4.3 Deepening rational use of wetland resources

In recent years, wetland management departments at all levels in Hubei Province have increased the pilot demonstration of the rational use of wetland resources, provided guidance on various wetland use activities, and actively carried out ecotourism, ecological agriculture, ecological education, and nature experience activities in line with the requirements of wetland conservation.

Dajiu Lake National Wetland Park in Shennongjia restores wetlands through implementation of measures such as returning land for farming to wetlands, migration and relocation, and ecological water replenishment, and actively develops ecotourism under the premise of protecting resources. In 2018, the admission revenue of Dajiu Lake Wetland Park was 35 million yuan, the income of transfer in the park 20 million yuan, and the total comprehensive tourism revenue reached 350 million yuan.

Chilong Lake National Wetland Park in Qichun, Hubei actively carries out nature education and was identified as a demonstration base for wetlands propaganda and education in Hubei Province in 2015. The year of 2020 saw the creation of the education brand of Chilong Lake Nature Classroom, which combines the theme of B&B and a wetland experience, nature conservation, sightseeing and leisure in one. A nature classroom with a construction area of 900 square meters has been built, which can meet the needs of group activities of about 40 to 50 people. At the same time, a number of popular science books have been compiled and published such as *Illustrated Handbook of Wildlife in Chilong Lake National Wetland Park, Ecological and Natural Observation Manual of Chilong Lake National Wetland Park and Illustrated Handbook and User Guide of Aquatic Medicinal Plants in Qichun*, and the book *Color Atlas of Plants in Chilong Lake National Wetland Park* won the third prize of Hubei Provincial Science and Technology Progress. Since the second half of 2018, 80,000 primary and secondary school students have participated in study tours each year, creating a cumulative economic benefit of about 60 million yuan.

In recent years, Hubei has been vigorously building small and micro wetlands, and actively creating a series of "wetland conservation +" models such as wetlands + water management model, wetlands + construction of water conservancy project model, wetlands + ecological tourism model, wetlands + rural revitalization model, and wetlands into the city + comprehensive management of the water environment model. At the same time, it aims to achieve synergistic symbiosis of wetland conservation and industrial economy based on development. The rural small and micro wetlands in Xujiazhuang village, Yuanan, Hubei are small and micro wetlands of ecological agriculture type. The small and micro wetlands give full play to their natural advantages and build ecological small and micro wetlands with the characteristics of "one village, one product". Through the construction of beautiful country roads, a natural education base has been built as an integrated one of ecological tourism, science education, and teaching practice. At the same time three-dimensional development has been realized in agriculture through the shrimp-rice continuous cropping, and farmers' income has increased. The shrimp-rice fields of the same unit area in small and micro wetlands procude more than 300kg of high-quality rice and 200kg of crawfish a year, average income of one mu of more than 4,500 yuan, an increase of more than 3,500 yuan than a single planting of rice, driving 27 households and 57 surrounding farmers (poor households population) out of poverty and increasing their wealth.

2.4.4 Heightened public awareness of wetland conservation

Every year Hubei Province takes publicity activities such as World Wetlands Day and Bird Loving Week as an opportunity to increase the wetland conservation publicity, and continue to raise awareness of wetland conservation. 2022 is the first year of implementation of the *Wetland Conservation Law* and the Standing Committee of Provincial Committee and Standing Committee of the Provincial NPC studied and deployed the implementation of the publicity work of Wetland Conservation Law; the Provincial People's Congress and the Provincial Government held a video conference on the implementation of the *Wetland Conservation Law*; the Provincial People's Congress and the Provincial Forestry Administration gave a press briefing on the implementation of the *Wetland Conservation Law*; the Provincial government issued opinions on the implementation of the *Wetland Conservation Law*. Forestry Administration of Hubei Province, Hubei Media Group, Hubei Science and Technology Press published the promotional film *Beautiful Wetlands Nourish Hubei*, and the themed picture album The Beauty of Wetlands in Hubei ("one film and one album") , organized national and provincial mainstream media to cover the achievements of wetland ecological conservation and restoration of the province in batches. The Publicity Department of Hubei Provincial Committee and the provincial river and lake chief scheme office jointly issued the *Implementation Plan of Comprehensive Promotion of Six Into of River and Lake Chief Scheme*, to promote the river and lake chief scheme into the party school, government organs, enterprises, communities, rural areas and schools and universities so that the awareness of wetland conservation of the whole community can be heightened significantly.

2.4.5 Successful bid to host the 14th Conference of the Contracting Parties to the Ramsar Convention

In February 2019, Hubei Province officially launched the work related to hosting the 14th Conference of the Contracting Parties (COP 14) to the Ramsar Convention, and through the active efforts of many parties, at the 57th session of the Standing Committee of the Ramsar Convention in 2019, the topic of Wuhan, Hubei Province hosting the 14th Conference of the Contracting Parties to the Ramsar Convention was considered and adopted, making it official that the COP 14 to the Ramsar Convention will be held in Wuhan, Hubei in November 2022.

第三篇

湖北湿地保护利用战略展望

CHAPTER 3

OUTLOOK OF WETLAND
CONSERVATION AND UTILIZATION
STRATEGY IN HUBEI

丹江口水库　　Danjiangkou Reservoir

　　湖北是生态大省，是长江径流里程最长的省份，是三峡大坝和南水北调中线工程核心水源区丹江口水库所在地，鄂西武陵山区是同纬度生物多样性最丰富的地区，神农架是北半球中纬度地区保存最完好的物种基因库之一，大别山、幕阜山是长江中下游重要的水源涵养地。湖北湿地在涵养水源、净化水质、维护生物多样性、蓄洪防旱、调节气候和固碳储碳等方面发挥着重要作用，在助推长江大保护、维护长江流域生态安全，确保"一库清水北送""一江清水东流"等方面发挥着基础性、战略性作用。

　　在长江大保护的国家战略背景下，湖北坚持走生态优先、绿色发展的道路，继续统筹推进长江生态环境高水平保护和经济社会高质量发展。基于此，湖北省湿地保护利用战略以全面提升湖北省湿地生态安全屏障质量、促进湿地生态系统良性循环和永续利用为总体目标，以《湿地保护法》实施为契机，全方位推动湿地保护管理；强化法治化建设，提升湖北智慧湿地综合管理能力，创新管理机制，探索跨区域生态补偿机制，稳步推进湿地可持续利用，不断满足新时期建设生态文明和美丽湖北对湿地生态资源的多样化需求，为长江经济带绿色发展提供生态保障。

一、以《湿地保护法》实施为契机，全方位推动湖北湿地保护管理

1. 加强湿地普法宣传，提高法治意识

将《湿地保护法》纳入普法重要内容，组织全省各级领导干部专题学习，提高法治意识和法治能力，让广大干部职工全面掌握《湿地保护法》的核心要义和基本要求。广泛开展使法律"进机关、进乡村、进社区、进学校、进企业、进单位"宣传活动，组织省内主要媒体持续性、全方位开展释法宣传，面向基层、面向群众广泛宣传《湿地保护法》的新规定、新要求，推动《湿地保护法》不断深入人心，形成学法、懂法、守法、用法的良好氛围。

2. 构建分级体系，完善制度和机制建设

加快建立湿地分级管理体系，编制湿地名录，加快构建湿地分级保护体系。根据国家要求，出台湿地面积总量管控目标方案，并将总量管控目标纳入湿地保护目标责任制。指导全省各地对照《湿地保护法》，推进地方性配套法规、政府规章和规范性文件等制度修订，完善湿地保护法治体系。开展湿地生态补偿、湿地资源调查评价、重要湿地动态监测评估和预警。召开湿地保护联席会议，建立湿地保护协作和信息通报等制度。

3. 科学规划，全面保护修复湿地

加快编制出台《湖北省湿地保护规划》及各地方的湿地保护规划，明确全省湿地保护修复的主要目标、空间布局和重点任务。强化湿地分级分类管理，切实加强国家和省级重要湿地保护，抓好重要湿地修复方案编制审批以及湿地修复验收、后期管理、修复效果评估，对重要湿地修复实行全过程监管。抓好湿地类型的国家公园、湿地自然保护区和湿地公园、湿地保护小区的建设管理。

4. 实践创新，打造新型"湿地保护+"模式

科学谋划，系统治理，在湿地保护修复过程中创建一系列"湿地保护+"模式，如"湿地+水体治理"模式、"湿地+水利工程建设"模式、"湿地+生态旅游"模式、"湿地+乡村振兴"模式、"湿地入城+水环境综合治理"模式等，实践创新、因地制宜，打造新型"湿地保护+"模式。

5. 立足发展，实现湿地保护与产业经济协同共生

坚持保护与发展并行，打造湿地生态旅游、湿地民宿体验、湿地自然教育、湿地文化与农耕文化相结合的乡村旅游等一批产业项目，实现湿地保护的经济效益、社会效益和生态效益协同共生。

6. 加强监督管理，落实河湖长制和林长制的湿地保护责任

认真落实湿地保护目标责任制，将湿地保护纳入湖长制、林长制考核评价体系，进一步压实地方政府湿地保护的主体责任，切实做好湖北省重点湖泊退垸还湖（还湿）工作。加大检查指导力度，督导各级林业部门和湿地公园认真履职尽责，严格落实守土有责要求，强化安全巡查，强化管护常态化，确保湿地资源不遭受破坏。处理好湿地保护与项目建设的关系，加大对占用湿地工程的监管，严格实行用途管制。

"十四五"时期，湖北省的林业工作将按照"四屏一系统"，即"鄂西南武陵山、鄂西北秦巴山、鄂东北大别山、鄂东南幕阜山四大森林生态屏障和鄂中平原湿地生态系统"，优化发展空间布局，维护生物多样性和生物安全，筑牢"三江四

湖北长江新螺段白鱀豚国家级自然保护区
Hubei Albino Dolphins National Nature Reserve in Xinluo Section of Yangtze River

武汉安山国家湿地公园　Wuhan Anshan National Wetland Park

圆穗蓼花海（神农架大九湖）　flower sea of Polygonum macrophyllum (Dajiu Lake in Shennongjia)

武汉东湖湿地　East Lake wetland in Wuhan

鹤峰县走马镇木耳山万亩茶树
thousands of mu of tea trees of Mu'er Mountain in Zouma town of Hefeng county

屏千湖一平原"生态安全格局。

湖北省将推进实施长江、汉江、清江及平原湖区生态廊道，森林城市建设，乡村绿化，重要生态系统保护和修复，湿地资源保护，自然保护地和野生动植物保护，森林质量提升等重大生态工程，不断提升生态系统质量和稳定性。

7. 更新理念，持续发力，湖北湿地保护上新台阶

湖北省坚持习近平生态文明思想，推动湿地保护再上新台阶；坚持湿地保护与合理利用并重，增强民众幸福感和获得感；坚持科学技术引领驱动，推动湿地保护创新发展；坚持健全规章制度，守护好湿地保护成果；坚持湿地保护的持续投入，争取更大的成绩。

二、强化法治化建设，加快颁布《湖北省湿地保护条例》

湖北省地处长江中游，纵跨江汉两大水系，境内江河纵横交错，湖泊星罗棋布，享有"千湖之省""鱼米之乡"的美誉，湿地资源极为丰富。建议尽快制定并出台相关法规制度。

湿地保护法要求"省、自治区、直辖市和设区的市、自治州可以根据本地实际，制定湿地保护具体办法"。考虑到各地情况存在差异，湿地保护法为地方制定具体办法留下空间。授权各地在本法和相关法律规定的范围内制定适合当地实际的湿地保护具体办法，有利于在本法规定的基础上具体落实本法实施。

湖北省因地制宜，着手推进制订湖北省湿地保护条例，从立法上来保护湖北省丰富的湿地资源，强化对湿地利用的监督与管理，完善我国自上而下的湿地法律体系，进而推动国家层面湿地保护立法的实现。湿地的定义、管理体制、基本制度、特殊规定、法律责任等方面是湖北省湿地保护条例和配套制度建设的重点内容。

三、完善智慧湿地系统，综合管理能力上新台阶

湖北省湿地保护工作在全面深化生态文明体制改革中扮演着十分重要的角色，为切实提升湖北省湿地保护管理水平，在湖北省"智慧林业"建设的部署下，进一步建设和完善湖北省"智慧湿地"综合管理平台。以全省湿地保护管理工作需求为导向，以提高全省湿地资源保护管理能力为宗旨，确立"动态监测体系、信

武汉杜公湖国家湿地公园　Wuhan Dugong Lake National Wetland Park

息共享机制、业务协同管理"等目标，结合计算机网络、物联网、大数据管理和分析等技术，实现对湿地资源的管理、监测、变化预测分析等，为湿地保护和合理利用提供辅助决策支持服务。

通过建设完善湖北省"智慧湿地"综合信息平台，有效提高湖北省各级湿地保护管理部门的湿地保护管理水平，有助于保护湿地资源和改善湿地生态环境。

同时，省级和各地级"智慧湿地"系统的建设和运行管理，也可用于湿地宣教渠道和方式的扩充，宣教方式更加丰富多样，宣教广度覆盖了社会所有群体，有助于营造全民热爱湿地和参与保护的良好社会氛围。

四、持续推进国际合作，促进国际重要湿地和湿地城市建设

2022年是中国加入《湿地公约》30周年。30年来，中国大力推进湿地保护修复，湿地生态状况得到持续改善。中国已迅速成为湿地保护领域的领导者，为全球生态治理贡献中国智慧和中国方案。

在中国方案的贡献中，湖北省湿地保护和科学管理等方面表现很突出。截至2021年底，全省已建成国际重要湿地4个（洪湖、沉湖、网湖、大九湖）、国家

武汉紫阳湖公园　*Ziyang Lake Park in Wuhan*

情迷大九湖　in love with Dajiu Lake

重要湿地8个、省级重要湿地54个；2022年6月8日，《湿地公约》官方网站公布了全球第二批国际湿地城市名单，湖北省武汉市获此殊荣，在湖北省和武汉市政府的领导、支持和公众积极参与下，武汉市湿地保护与修复工作取得明显成效，得到了国际社会的认可，武汉市生态文明建设迈上了新台阶。

湖北省将以《湿地公约》第十四届缔约方大会在武汉召开为契机，进一步加强湿地保护的国际合作。加强与全球环境基金（GEF）、联合国开发计划署（UNDP）、世界自然基金会（WWF）等组织的合作，进一步提高湿地保护科技含量，全面提高全省湿地保护与建设的能力和水平。通过强化合作与交流，积极探索湿地保护与合理利用的有效模式，引导人们转变生产生活方式，促进改善湿地所在地民生，提高可持续发展能力。

通过国际合作，推进国际重要湿地和湿地城市建设示范，着力加强国际重要湿地保护和武汉国际湿地城市建设管理，提高管理水平；进一步完善湖北湿地保护体系，积极申报更多国际湿地城市和国家（国际）重要湿地。

Chapter 3

Outlook of Wetland Conservation and Utilization Strategy in Hubei

Hubei is a province featuring ecology with the longest runoff mileage of the Yangtze River, and the location of Three Gorges Reservoir Area and Danjiangkou, the core water source area of the medium route project of the South-to-North Water Diversion. The Wuling Mountains in the western Hubei is the richest area in biodiversity at the same latitude; the Shennongjia region is the best preserved species gene pool in the mid-latitude region of the northern hemisphere; and the Dabie Mountains and the Mufu Mountains are important water-conserving areas in the middle and lower reaches of the Yangtze River. The wetlands in Hubei play an important role in water conservation, water purification, maintenance of biodiversity, flood storage and drought prevention, climate regulation and carbon sequestration and storage, and a fundamental role in promoting the well-coordinated environmental conservation of the Yangtze River, maintaining the ecological security of the Yangtze River basin, and ensuring a fundamental and strategic role in "a reservoir of clean water is sent north" and "a river of clean water flows east".

Against the backdrop of the national strategy of well-coordinated environmental conservation of the Yangtze River, Hubei will adhere to the path of ecological priority and green development, and continue to promote the high-level conservation of the ecological environment of the Yangtze River and high-quality economic and social development in an integrated manner. Based on this path, the strategy of wetland conservation and utilization in Hubei should take comprehensively improving the barrier quality of wetland ecological security and promoting the benign cycle of wetland ecosystems and sustainable use as the overall goal, the implementation of the Wetland Conservation Law as an opportunity to promote the conservation and management of wetlands in Hubei in all aspects. Hubei also strengthens its legal system, improves the comprehensive management capacity of intelligent wetlands, innovates management mechanisms, explores mechanisms of cross-regional ecological compensation, steadily promotes sustainable use of wetlands, continuously meeting the diverse needs in the new era of building ecological civilization and a beautiful Hubei for wetland ecological resources, and providing ecological security for the green development of the Yangtze River Economic Belt.

Chapter 3
Outlook of Wetland Conservation and Utilization Strategy in Hubei

3.1 Take the implementation of the Wetland Conservation Law as an opportunity to promote the conservation and management of wetlands in Hubei in all aspects

3.1.1 Increase publicity of wetland laws, raise awareness of the rule of law

Wetland Conservation Law should be included as important content into the publicity of laws, and the leaders and cadres at all levels of the province should take project-based learning to improve their awareness of the rule of law and capabilities to rule based on laws so that the majority of cadres and workers can fully grasp the core essence and basic requirements of the Wetland Conservation Law. Hubei widely carries out publicity activities to make the law go "into the organs, into the countryside, into the community, into the schools, into the enterprises, into the units". It organizes the main media in the province and encourages them to continue to comprehensively publizice the law for the grassroots and the masses so that they can understand the new provisions and requirements of the Wetland Conservation Law and advocate it with all their heart, creating a favorable atmosphere of learning the law, understanding the law, abiding the law, and using the law.

3.1.2 Build a grading system, improve the construction of system and mechanism

Hubei should accelerate the establishment of wetland grading management system and prepare list of wetlands in order to speed up the construction of graded conservation system of wetlands. According to the national requirements, the program of the control target of total wetland area should be introduced, and the target should be included into the accountability system of wetland conservation target. The province should be guided to formulate and revise local supporting legislation, government regulations and regulatory documents based on the *Wetland Conservation Law* in order to improve the rule of law system for wetland conservation. Ecological compensation for wetlands, survey and evaluation of wetland resources, and dynamic monitoring and assessment and early warning of important wetlands should be carried out. Joint meetings on wetland conservation should be convened and systems of collaboration and information notification on wetland conservation should be established.

3.1.3 Plan in a scientific manner, comprehensively conserve and restore wetlands

The preparation and introduction of the *Wetland Conservation Plan in Hubei Province* and local wetland conservation plans should be accelerated to clarify the main objectives, spatial layout and key tasks of wetland conservation and restoration in the province. The classification and grading of wetland management should be strengthened to effectively enhance the conservation of wetlands of national and provincial importance. High attention should be paid on the preparation and approval of restoration programs of important wetlands, as well as acceptance of wetland restoration, post-management, assessment of restoration effect, and the restoration of important wetlands should be supervised during the whole process. Great efforts should be made to ensure the construction and management of national parks of wetland type, wetland nature reserves and wetland parks, and wetland conservation areas.

3.1.4 Practice innovation, create a new model of "wetland conservation +"

A series of models of "wetland conservation +" should be created in the process of wetland conservation and restoration through scientific planning and systematic management, such as a model of wetlands + water management, wetlands + construction of water conservancy, wetlands + ecological tourism, wetlands + rural revitalization, and wetlands into the city + comprehensive management of water environment. Hubei will practice and innovate according to local conditions to create new models of "wetland conservation +".

3.1.5 Focus on development, achieve synergy between wetland conservation and industrial economy

Hubei adheres to conservation while pursuing development, and creates a number of industrial projects such as wetland ecotourism, wetland B&B experience, wetland nature education, and rural tourism combining wetland culture and farming culture, to achieve synergy between the economic benefits of wetland conservation, social benefits and ecological benefits.

3.1.6 Strengthen supervision and management, bear responsibilities of wetland conservation of river and lake chief scheme and forest chief scheme

Hubei earnestly implements the responsibility system of wetland conservation target and includes wetland conservation into the assessment and evaluation system of lake chief scheme and forest chief scheme to see main responsibilites of local government are fulfilled in wetland conservation, and make solid efforts to return land for levees to the lakes (and wetlands) of key lakes in Hubei Province. Hubei increases inspection and guidance by making sure that forestry departments at all levels and wetland parks diligently perform their duties and responsibilities, strictly implementing the requirements of land guarding and strengthening security inspections and the normalization of management and care, in order to to ensure that wetland resources do not suffer damage. Hubei handles the relationship between wetland conservation and project construction, increases the supervision of the project to occupy wetlands, and strictly controls the use of wetlands.

Chapter 3
Outlook of Wetland Conservation and Utilization Strategy in Hubei

During the 14th Five-Year Plan period, Hubei forestry will follow the "four barriers and one system", i.e. "four forest ecological barriers in the Wuling Mountains of southwest Hubei, the Qinba Mountains of northwest Hubei, the Dabie Mountains of northeast Hubei, the Mufu Mountains of southeast Hubei, and wetland ecosystem of central Hubei Plain" to optimize the spatial layout of development, maintain biodiversity and biosecurity, and build a firm ecological security pattern of "three rivers and four barriers, a thousand lakes and a plain" proposed by the Provincial Party Committee.

Hubei Province will promote the implementation of the ecological corridors of the Yangtze River, Hanjiang River and Qingjiang River and the plain lake area, construction of forest city, rural greening, conservation and restoration of important ecosystems, wetland resources conservation, nature reserves and wildlife conservation, forest quality improvement and other major ecological projects, and constantly improve the quality and stability of the ecosystem.

3.1.7 Update concept with sustained efforts, lift wetland conservation in Hubei up to a new level

Hubei Province adheres to Xi Jinping's thought of ecological civilization and lifts wetland conservation up to a new level. It attches equal importance to wetland conservation and rational use of wetlands, makes people have a greater sense of happiness and benefit, adheres to lead the drive with science and technology, promotes the innovative development of wetland conservation and perfect rules and regulations, guards the achievements of wetland conservation, and continuously invests in wetland conservation for greater achievements.

3.2 Strengthen legal system and speed up the promulgation of Regulations on Wetland Conservation in Hubei Province

Hubei Province is located in the middle reaches of the Yangtze River, straddling the two major water systems. With rivers and lakes crisscrossing in the territory and abundant wetlands resources, Hubei enjoys the reputation of "province of a thousand lakes" and "home of fish and rice". It is recommended that the *Regulations on Wetland Conservation in Hubei Province* be enacted and introduced as soon as possible to protect the rich wetland resources in Hubei Province from legislation, strengthen the supervision and management of use of wetlands, improve the top-down wetland legal system in China, and then promote the realization of legislation of wetland conservation at the national level.

The *Wetland Conservation Law* requires that "provinces, autonomous regions, municipalities, cities divided into districts, and autonomous prefectures can develop specific measures to protect wetlands according to local situations." Since there are differences in local conditions, the *Wetland Conservation Law* leaves room for local development of specific measures and authorizes local governments to develop specific measures suitable for local realities to protect wetlands within the scope of this law and related laws, which is conducive to the specific implementation of this law on the basis of the provisions of this law.

Hubei Province should formulate regulations on wetland conservation in line with the local conditions in the province. The definition of wetlands, management system, the basic system, special provisions and legal liability are the focus of regulations on wetland conservation in Hubei Province and the construction of its supporting systems.

3.3 Improve intelligent wetland system to lift comprehensively management capacity up to a new level

Wetland conservation in Hubei Province plays a very important role in the comprehensive deepening of the system reform of ecological civilization. In order to effectively lift the level of wetland conservation and management in Hubei Province, Hubei should further building and improving the comprehensive management platform of "intelligent wetlands" of the province under the deployment of the construction of "intelligent forestry". The province should, with the needs of wetland conservation and management as the guide and improvement of the wetland resources conservation and management capabilities as the purpose, identify the objectives of establishing "dynamic monitoring system, information sharing mechanism, business collaboration management", achieve the management, monitoring, prediction and analysis of changes of wetland resources, and provide supportive services of auxiliary decision-making for wetland conservation and rational use with the help of technoligies such as computer networks, Internet of Things, big data management and analysis.

Through the construction and improvement of the comprehensive information platform of "intelligent wetland" in the province, Hubei effectively improves the level of wetland conservation and management of departments at all levels of wetland conservation and management in the province, helping to conserve wetland resources and improve the ecological environment of wetlands.

At the same time, the construction and operation management of the provincial and local-level intelligent wetland systems can also be conducive to the expansion of wetland education channels and methods, making them richer and more diverse and cover all groups in society, as well as to creating a favorable social atmosphere for all people to love wetlands and participate in conservation.

3.4 Promote continously international cooperation to advance construction of wetlands of international importance and wetland cities

The year 2022 marks the 30th anniversary of China's accession to the *Convention of Wetlands of International Importance Especially as Waterfowl Habitats* (*Ramsar Convention*). Over the past 30 years, China has been vigorously promoting wetland conservation and restoration, and the ecological condition of wetlands has been continuously improving. China has rapidly become a leader in the field of wetland conservation, contributing China's wisdom and China's solutions to global ecological governance.

Among China's solutions and contributions, Hubei Province is prominent in wetland conservation and scientific management. By the end of 2021, the province has built four wetlands of international importance (Honghu Lake, Chenhu Lake, Wanghu Lake and Dajiu Lake), eight wetlands of national importance and 54 wetlands of provincial importance, and the wetlands of national importance and international importance rank the 1st and 2nd in China respectively. on June 8, 2022, Ramsar Convention Bureau announced the list of the second batch of international wetland cities in the world, and Wuhan, Hubei Province was among the prize winners. Under the leadership of Hubei Province and the city government of Wuhan, support and active participation from the public, wetland conservation and restoration work in Wuhan, Hubei has achieved significant results, gained recognition of the international community, and lifted construction of ecological civilization up to a new level in Wuhan.

Hubei Province should take the 14th Conference of the Contracting Parties to the Ramsar Convention held in Wuhan as an opportunity to further strengthen international cooperation in wetland conservation. It should strengthen cooperation with organizations such as the Global Environment Facility (GEF), the United Nations Development Programme (UNDP) and the World Wide Fund for Nature to further improve the scientific and technological level of wetland conservation, and comprehensively improve the capacity and level of wetland conservation and construction of the province. Through enhanced cooperation and exchange, it should actively explore effective models of wetland conservation and rational use of wetlands, guide people to change their way of production and lifestyle, improve the livelihood of people in wetland areas, and enhance sustainable development.

Through international cooperation, Hubei promotes the demonstration of the construction of wetlands of international importance and wetland cities, focuses on strengthening the conservation of wetlands of international importance and the construction and management of Wuhan as an International Wetland City, and improves the management level. It also further improves the wetland conservation system in the province, and actively declares international wetland cities and wetlands of national (international) importance.

第四篇

湖北湿地保护管理典型案例

CHAPTER 4

TYPICAL CASES OF WETLAND CONSERVATION AND MANAGEMENT IN HUBEI

武汉长江大桥　　Yangtze River Bridge in Wuhan

　　近年来，湖北省认真贯彻习近平生态文明思想，牢固树立和践行绿水青山就是金山银山理念，统筹山水林田湖草系统治理；认真落实中央和湖北省委、省政府关于湿地保护工作的决策部署，切实履行湿地保护管理职责，持续推进湖北湿地保护修复，取得了阶段性明显成效。

　　在全面展示湖北省生态文明建设实践成果、深入介绍湖北重要湿地保护成就

的基础上，我们总结了湖北省在湿地保护与恢复、制度建设与创新管理、合理利用与社区共建、科普宣教与科研监测等方面的探索、创新和示范经验。依据湿地生态重要性、重要湿地功能、保护，科学管理及科普科研等方面，选取了湖北典型湿地保护、修复及综合管理等具有代表性的案例和创新模式，探讨湖北湿地保护管理的创新创先之实践，同时展示湖北湿地生态之美、人与自然和谐之美。

一、武汉国际湿地城市——城市湿地保护与综合管理

1. "百湖之市"的武汉

武汉市,号称"百湖之市""湿地之城",中国最大的河流长江与它最大的支流汉江在市中心交汇,淡水资源得天独厚。全市境内江河纵横,河港沟渠交织,湖泊库塘星罗棋布,长江、汉江交汇于市境中央,且接纳南北支流入汇,境内河流5千米以上的有165条,现有水面2117.6平方千米。市内现有大小湖泊166个,众多大小湖泊镶嵌在大江两侧,形成湖沼水网,知名的有东湖、南湖、汤逊湖、金银湖、沙湖等,构成了极具特色的滨江、滨湖水生态环境。在正常水位时,湖泊水面面积居中国城市首位,是我国内陆湿地资源较丰富的特大城市之一,也是全球同纬度和长江中下游湖泊湿地的典型代表地区。

武汉湿地面积16.2万公顷,占全市国土面积的18.9%。其中,自然湿地11.3万公顷,占湿地面积的69%;人工湿地4.97万公顷,占湿地面积的31%,湿地资源居全球内陆城市前三位。全市湿地保护率为53.19%。武汉湿地生物多样性丰富,栖息着413种野生动物,其中国家重点保护野生动物38种,包括东方白鹳、黑鹳等10种国家一级保护动物,灰鹤(*Grus grus*)、白琵鹭等28种国家二级保护动物。

武汉东湖　East Lake in Wuhan

东方白鹳　*Ciconia boyciana*

湿地上分布有维管束植物408种,其中国家二级保护野生植物有野大豆(*Glycine soja*)、野莲、野菱、粗梗水蕨、水蕨等5种。

湿地作为武汉城市之肾,在维持生态环境稳定,特别是在减轻洪水威胁、弱化城市热岛效应和控制环境污染等方面发挥着重要作用。武汉要实现城市可持续发展,在中部发展战略中扮演好中心城市的角色,必须合理利用和保护好湿地这一优势资源,为城市建设服务。

2. 武汉湿地保护地体系建设

自1994年设立沉湖湿地自然保护区开始,武汉湿地保护已有近30年历程。武汉不断完善湿地保护法律法规和制度建设,改革创新湿地保护体制机制,实施江滩改造、一湖一景、大东湖生态水网等重大工程,推广应用先进的湿地保护技术,动员市民参与湿地保护,修复湿地生态环境。目前,武汉已形成以湖泊保护为特色的湿地保护法规体系,以自然保护区与湿地公园为核心的湿地保护体系,保护地数量在全国城市中首屈一指。现有1处国际重要湿地——蔡甸沉湖,2个省级自然保护区——蔡甸沉湖、新洲涨渡湖,5个国家湿地公园——东湖、安山、后官湖、杜公湖、藏龙岛,4个省级湿地公园——蔡甸索子长河和桐湖、江夏潴洋海、黄陂木兰花溪,3个市级湿地自然保护区——江夏上涉湖、黄陂草湖、武湖,还有1处

武汉金银湖国家城市湿地公园
Wuhan Jinyin Lake National Urban Wetland Park

武汉安山国家湿地公园
Wuhan Anshan National Wetland Park

武汉后官湖国家湿地公园
HouHuanHu National Wetland Park

国家城市湿地公园——金银湖。

结合武汉地区候鸟迁徙通道以及国家重点保护野生动物集中取水取食等区域，将黄陂祁家湾（青头潜鸭）、洪山天兴洲（黑鹳）、东西湖府河柏泉段（天鹅湖）等3处列为我市野生动物重点保护栖息地。

3. 武汉湿地保护管理制度和创新机制

武汉是全国最早将湿地保护与修复纳入法制化轨道的城市。1999年，武汉市政府就出台了《武汉市自然山体湖泊保护办法》，将湖泊湿地保护与山体保护并重。2007年，出台《武汉市湿地保护总体规划》，将湿地保护纳入城市总体规划。2010年3月，武汉在全国副省级城市中率先完成湿地立法，颁布了《武汉市湿地自然保护区条例》，划定了一批湿地公园和自然保护区，形成了湿地保护的"武汉样本"。2013年10月，武汉市在全国第一个推出湿地生态补偿机制——《武汉市湿地自然保护区生态补偿暂行办法》，每年投入8亿多元进行生态补偿，建立湿地保护的生态补偿机制。武汉市启动了"智慧湿地"项目建设，并在全国率先编制《武汉市小微湿地保护与修复指南》。同时建立了柏泉、杜公湖、汤湖等第一批小微湿地

府河湿地公园　Fuhe River Wetland Park

黄陂梅店水库　Meidian reservoir in Huangpi

生态修复示范点，加大湿地自然保护区及天兴洲、祁家湾、府河柏泉段等野生鸟类栖息地保护力度，开展沉湖国际重要湿地和天兴洲生物多样性、长江干流重点保护陆生野生动植物资源、越冬水鸟调查。

通过构建多重湿地保护体系，武汉已建成全国最大的长江滨江江滩景观，东湖也获评"长江经济带最美湖泊"；建立河湖长制，以流域为着眼点统筹上下游左右岸；全市166个湖泊组成国内最大城市湖泊生态湿地群；给165条河流留足滨水空间和绿化缓冲带。

经过逐年积累，武汉已经形成"政府主导、部门协同、社会参与"的湿地保护合力，30余个非政府组织机构、20万志愿者助力湿地生态保护。此外，依托湿地资源，全市建立至少12处宣教场馆、69处科普基地，还打造了中国履行《湿地公约》30周年成果展示馆。全市围绕世界湿地日、世界野生动植物日、爱鸟周等开展了一系列主题宣传,还出版了《寻梦守泽 "湿"意江城》,组织拍摄纪录片《湿地上的城市》,多样展示武汉湿地保护成就,讲述武汉湿地故事。

4. 实施重要湿地保护和恢复工程

近年来,武汉通过湿地岸线修复、河湖水系连通、野生动物栖息地恢复等举措,持续开展两江四岸环境整治、大东湖生态水网建设、六湖连通、四水共治、海绵城市试点等一系列湿地保护与生态修复工作,成效显著。

沉湖国际重要湿地通过疏通水系、保育原生植被、构建多样化鸟类栖息地、建立智慧湿地监管模式等,修复沉湖湿地生态系统,提升保护管理能力,沉湖湿地生态环境得到持续改善。

东湖绿岛驿站改造提升工程。

涨渡湖水上森林新造池杉林、种植水生植物、修建护鸟隔离栅栏、观鸟栈道、水域岸线环境综合整治。

杜公湖湿地保护和恢复工程:退耕还湿、退居还湿、退建还湿、沼泽湿地恢复等,生态补水工程、湖泊整治工程、植被恢复与生境改善等。

水雉　*Hydrophasianus chirurgus*

涨渡湖池杉林　Chishan forest of Zhangdu Lake

长江天鹅洲故道湿地　　wetlands in Tian-e-zhou Oxbow of the Yangtze River

二、天鹅洲长江故道湿地——珍稀物种保护地管理探索

1. 长江中游故道群湿地的特殊性和典型性

荆江是长江在中游冲积平原上的一段河道，受地形、地质、水文等因素的影响，在水深流急、崩岸频繁的长江荆州河段，河道自然变迁以及人为的影响导致本来弯曲的河道裁弯取直，上下口逐渐淤塞封闭而形成长江故道（牛轭湖）。长江故道的洲滩形成了大面积的湿地。天鹅洲故道是1972年长江自然裁弯形成，原为通江故道，1998年修建沙滩子围堤与长江阻隔，仅通过下游天鹅洲闸与长江相通。

天鹅洲长江故道湿地范围内拥有湖北石首麋鹿国家级自然保护区和湖北长江天鹅洲白鱀豚国家级自然保护区，分别以麋鹿和长江江豚以及其栖息地为保护对象。此外，湿地区内还分布着世界自然保护联盟（IUCN）濒危物种：东方白鹳；易危物种：鸿雁（*Anser cygnoides*）、小白额雁（*Anser erythropus*）、大滨鹬（*Calidris tenuirostris*）、黄胸鹀（*Emberiza aureola*）；极危物种：青头潜鸭；近危物种：红脚隼（*Falco amurensis*）、日本鹌鹑（*Coturnix japonica*）、长嘴半蹼鹬（*Limnodromus scolopaceus*）、黑尾塍鹬（*Limosa limosa*）、白腰杓鹬（*Numenius arquata*）、白颈鸦

（*Corvus pectoralis*）、红颈苇鹀（*Emberiza yessoensis*）。天鹅洲长江故道区湿地生态系统典型，生态结构完整，功能独特，野生动植物资源丰富，对于典型独特湿地生态系统及物种多样性的保护具有重要意义。天鹅洲长江故道湿地集自然地理、珍稀动物以及自然湿地生态保护于一身，具有重要的科研、文化、生态及旅游价值。

2. 国家级自然保护区的建设和管理

（1）湖北石首麋鹿国家级自然保护区

湖北石首麋鹿国家级自然保护区，总面积1567公顷，主要任务是在麋鹿原生地恢复自然种群，并保护其赖以生栖的湿地生态环境。经过保护人员的多年努力，石首麋鹿国家级自然保护区麋鹿种群发展迅速，已由1993年10月和1994年11月分两批从北京南海子麋鹿苑引进的64头发展到如今的2500余头，并形成核心区、江南三合垸、小河杨波坦及湖南洞庭湖四个亚种群，且全部实现了自然放养，恢复了野生习性。

扩大麋鹿活动空间。为了解决人与鹿争地的矛盾，在各级党委政府的支持下，

麋鹿成群　Elaphurus davidianus in flocks

飞奔　galloping

鹿王风采　manner of the leader of flocks

哺乳　feeding

归　going home

保护区核心区、实验区 1000 公顷土地权属和缓冲区约 533 公顷土地使用权等到了落实，麋鹿活动空间逐步扩大。通过加强保护区内管控，确保了保护区范围内环境的安宁。

创造良好生存条件。一是食物保障。在加强天鹅洲生境保护前提下，建设麋鹿应急饲料基地。二是水源保障。通过建设提排水泵站和管网进行水体交换，确保水源清洁。三是隐蔽场所。保护区内拥有大面积的芦苇和旱柳林，能够为麋鹿的产崽、避风提供较好的隐蔽场所。

开展生物多样性监测。以麋鹿监测为重点，充分运用 3S（遥感技术、地理信息系统、全球定位系统）、无人机、红外相机等新技术、新设备开展监测。通过对麋鹿生活习性、活动规律、环境需求等方面的监测，也为保护、恢复、管理麋鹿栖息地提供科学依据。

做好栖息地管理。围绕麋鹿生活习性，因鹿施策开展栖息地管理；针对麋鹿的度汛需求，建设麋鹿度汛安全区人工牧草地保障区。

保护区办公园区　office parks in Nature Reserve

长江天鹅洲白鱀豚国家级自然保护区
Yangtze River Swan Oxbow White-flag Dolphin National Nature Reserve

强化疫源疫病防控和科研合作。保护区加强与科研单位和高等院校合作，找准保护区疫源疫病防控薄弱环节，建立健全麋鹿疫病预警和防控体系，以及科研监测平台。

（2）湖北长江天鹅洲白鱀豚国家级自然保护区

1992年，湖北长江天鹅洲白鱀豚国家级自然保护区成立，科研人员尝试利用天鹅洲长江故道，对长江濒危物种白鱀豚（现已灭绝）、长江江豚（*Neophocaena asiaeorientalis*）等进行迁地保护；自然保护区投放5头江豚试养至今，2021年经过普查，已发展成为数量为101头的种群，每年有6～8头小江豚的出生，为长江江豚自然种群恢复奠定了坚实基础。

天鹅洲长江故道成为长江江豚种群数量最多的保护区，石首市因此被称为"中国江豚之乡"。江豚迁地保护工作被国内外专家认定为对一种鲸类开展迁地保护唯一成功范例。在长江流域江豚种群总体数量急剧减少状况下，天鹅洲故道江豚的迁地保护工作给长江江豚的生存和保护带来了希望，也为其他长江珍稀濒危水生动物的保护提供了宝贵经验。2021年2月，经国务院批准，国家林草局、农业农村部公布新修订的《国家重点保护野生动物名录》，其中长江江豚升级为国家一级保护野生动物。江豚，这一长江旗舰物种，来自大自然的"微笑天使"，在人们的精心呵护下，畅游长江。

3. 故道群湿地与濒危物种迁地保护

迁地保护往往是对种群数量急剧下降的物种采取的非常重要而且必要的保护手段。天鹅洲故道湿地主要是用过建立迁地保护区，引进麋鹿、长江江豚等濒危珍稀物种，以达到对整个湿地生态系统的保护。

（1）长江江豚保护

经过30年的发展，天鹅洲保护区江豚保护工作取得成效，保护区被国内外专家一致认定为对一种鲸类动物开展迁地保护唯一成功的范例。到2021年，天鹅洲故道江豚种群数量已逾100头，且每年增长10%左右，已达到江豚种群环境容纳量，是目前长江流域范围内江豚种群数量唯一稳定增长的区域。

开展长江江豚迁地保护。在保护区长江89千米的石首江段，通过开展拆违打

长江江豚　*Neophocaena asiaeorientalis*

非、环境综合整治、生态修复等工作，现石首长江段常年栖息江豚20余头，且经常成群出现。

开展尝试网箱人工豢养江豚。保护区网箱豢养江豚和网箱江豚繁育技术分别获得国家发明专利。2020年7月，网箱繁育的第一头江豚"贝贝"在年满4岁之际通过野化训练已放归于天鹅洲故道。

科学推进江豚种群基因交换。为改善天鹅洲故道江豚种群基因结构，防止江豚种群基因衰退，从2015年开始，天鹅洲保护区先后向新螺段、湖北监利何王庙、安徽铜陵等保护区以及其他适宜的栖息地输出了24头江豚，同时从江西鄱阳湖迁入了8头江豚。按照《长江江豚拯救行动计划》，保护区后续将不断采取长江江豚输出迁入措施，改善江豚种群基因。

（2）麋鹿及其栖息地保护

通过有效的保护，保护区生物多样性得到了提高，麋鹿种群数量得以扩大。最新的综合科考显示，保护区高等植物由1997年的238种增加到了321种，陆生脊椎动物由1997年的90种增加到了320种，国家重点保护物种达到了56种，特别是近年来先后发现了黑鹳、东方白鹳、白鹤、白枕鹤等珍稀濒危鸟类。

重要保护对象麋鹿则从最初的64头发展到2500余头，并形成了核心保护区、江南三合垸、小河杨波坦及湖南洞庭湖四个亚种群，且全部实现了自然放养。保护区先后两次被国家七部委联合表彰为"全国自然保护区先进集体"，并连续两次在全国国家级自然保护区管理工作评估中被列为优秀等次。

三、神农架国家公园——湿地保护与管理创新探索

1.神农架国家公园生态定位与区域湿地特色

神农架是北半球中纬度地区保存最完好的常绿落叶阔叶混交林，具有自然环境独特，物种多样、古老、孑遗、稀有，水资源丰富等特点，有着独具特色的旅游资源和珍稀的动植物资源，素有"自然博物馆""物种基因库""孑遗动植物避难所"之美誉，是野生动植物生存繁衍的理想场所。神农架具有完好的森林生态系统，丰富的降水量，是长江流域重要的水源涵养区，在改善生态环境和维护生态平衡等方面发挥着重要作用。

大九湖湿地　Dajiu Lake wetland

神农架大九湖湿地位于湖北西北端,坐落于长江和汉江的分水岭上,西与重庆、陕西接壤,北与湖北十堰市为邻,东与恩施巴东县相连。大九湖湿地平均海拔1730米,被神农架群山所环绕,形成了独特的高山盆地,构成别具一格的盆地与山岳相间的自然景观,天然森林植被与湿地沼泽植被交错的迷人画卷。神农架大九湖是湿地生态系统和森林生态系统的完美结合,独特的湿地结构让大九湖湿地比其他湿地更具多样性。川金丝猴(*Rhinopithecus roxellana*)、金雕(*Aquila chrysaetos*)等珍稀动物在这里繁衍生息;以珙桐(*Davidia involucrata*)、鹅掌楸(*Liriodendron chinense*)、连香树(*Cercidiphyllum japonicum*)等为代表的孑遗植物在这特殊的自然环境中得到庇护,这样的湿地在中国也极为罕有,2013年更是被国际湿地公约组织列入国际重要湿地名录。

党的十八届三中全会做出了关于加强生态文明建设的决定,明确提出建立国家公园体制。2016年11月,神农架进入国家公园体制试点实施阶段。神农架国家公园体制试点区总面积1170平方千米,占神农架林区总面积的35.97%,试点区森林覆盖率高达96%以上。神农架国家公园地处中国地势第二阶梯的东部边缘,属于大巴山山脉东延的中高山区,是中国生物多样性保护优先区之一,且是长江与汉江的分水岭。神农架林区的堵河水系，流域面积791平方千米，汇入汉江后到达丹江口水库，水量占丹江口水库水量的20%。因此，神农架是长江经济带绿色发展的生态基石、南水北调中线工程重要的水源涵养区、三峡库区最大的天然绿色屏障，生态地位十分重要，关系着国家的生态安全及经济的可持续发展。

2. 神农架生态建设与湿地保护

神农架国家公园体制试点工作启动以来，践行习近平生态文明思想和"绿水青山就是金山银山"的理念，坚持生态保护第一、国家代表性、全民公益性的国家公园理念，以保护自然生态系统的原真性和完整性为目标，统筹山水林田湖草沙一体化保护，积极探索大九湖湿地保护和治理的新路径，使得大九湖生态环境得到明显改观、水源水质得到明显改善、湿地生态功能明显提升。

理顺管理体制，凝聚保护合力。神农架国家公园实行分区、分级、分类保护和管理，由单一森林生态系统保护，转变为山水林田湖草沙一体化保护。

坚持规范引领，强化法制保障。2018年5月《神农架国家公园保护条例》实施，实行"一园一法"管理；神农架国家公园还制定了产业准入正面和负面清单，形成了较为完善的法规和制度规范体系。

开展污染防治，全面落实河湖长制。坚持"一河一策""一湖一策"，及时解决制约水环境的重点、难点问题，累计关停9座小型水电站。

强化生态修复，保护重要湿地。为确保"一湖清水永续北送"，自2013年起，对大九湖湿地460户原居民实施了整体生态移民搬迁；针对湿地核心区划定了禁牧区，杜绝了大范围面源污染的可能。与此同时，还积极探索泥炭藓沼泽湿地保护和修复新思路。

建成立体网络，实现智能管理。园内已建成神农架国家公园信息管理中心、景区游客流量预警系统，运用卫星遥感、无人机、红外线摄像等监测手段，形成立体式保护网络，实现生物监测全覆盖。

3. 社区共建创新机制

神农架国家公园为巩固保护成果，探索可持续的保护管理与社区共建模式，自2016年11月以来，在国家公园辖区内5个乡镇开展社区共建共管、产业发展、基础设施建设、生态补偿及精准扶贫等方面累计共投入资金1.46亿元。

红豆杉 *Taxus*

泥炭藓 peat moss

红腹锦鸡 *Chrysolophus pictus*

川金丝猴 *Rhinopithecus roxellana*

小天鹅 *Cygnus columbianus*

143

第四篇
湖北湿地保护管理典型案例

大九湖湿地　Dajiu Lake wetland

建立社区共建共管联动机制。与辖区 5 个乡镇 25 个行政村签订社区共建共管合作协议，成立工作机构，建立健全联系工作机制，实现国家公园社区生态保护共责、产业发展共抓、能力提升共做、社会管理共建、改革成果共享的社区共建共管发展目标。

积极推进生态移民，开展生态补偿机制。神农架国家公园牢固树立共建共享理念，注重建立利益共享和协调发展机制。从实施生态移民搬迁以来，在生态移民补偿及安置区坪阡古镇建设共投资 9.76 亿元。大力实施以电代燃料试点工程，按照生态移民搬迁户每户每年 3000 元的标准实施以"电代燃料"补贴试点，改变以往的砍伐森林，燃烧木材的取暖方式，从而实现在解决居民冬季取暖的基础上最大限度地保护神农架森林资源的目的，最大限度降低居民生产生活用柴对生态资源的威胁。

坪阡古镇
Pingqian ancient town

四、洪湖国际重要湿地——湖泊保护与综合治理实践

1. 洪湖湿地生态区位与功能

"洪湖水，浪打浪，洪湖岸边是家乡……"，一曲经典老歌成为洪湖亮丽的名片。洪湖湿地位于湖北省东南部、长江中游北岸，江汉平原四湖流域下游，地跨洪湖和监利两市。2012年，"一湖一勘"确定的洪湖水面面积为308平方千米，是湖北省第一大湖泊，长江中下游大型湖泊之一。洪湖是一个以调蓄为主，兼具灌溉、渔业、航运和生物多样性保护等多种功能的重要湖泊。洪湖湿地于2008年被列入国际重要湿地名录，2014年12月晋升为国家级自然保护区。它还是中国湿地保护行动计划优先保护项目，被世界自然基金会确定的全球最重要的238个生态区之一。

洪湖湿地自然保护区风景优美，生物多样性丰富，湿地生态系统保持较好。据调查，保护区内共有维管束植物70科194属286种，其中国家重点保护植物4种。浮游植物143种。脊椎动物292种，其中陆生脊椎动物205种，包括鸟类173种、两栖类7种、爬行类12种、哺乳类13种，鱼类87种、浮游动物123种、底栖动

洪湖湿地　Honghu Lake wetland

荇菜　*Nymphoides peltata*

物 32 种。其中国家重点保护的脊椎动物有 40 种，属于国家一级保护的有东方白鹳、黑鹳、中华秋沙鸭、大鸨、白肩雕（*Aquila heliaca*）、白尾海雕等。

洪湖湿地属于内陆湿地和淡水水域生态系统类型的自然保护区，其主要保护对象是永久性淡水湖泊湿地生态系统、淡水资源及珍稀水禽，具有保护洪湖的淡水资源、生物多样性、维护湖泊生态系统结构和生态环境的功能，使洪湖湿地能够永久、持续、有效地服务区域经济、社会协调发展，促进区域内生态文明建设。

2. 洪湖湿地保护恢复与综合管理

洪湖地处长江中游江汉湖群，是荆楚大地重要的生态屏障。近年来，在湖北省委省政府，荆州市委市政府"共抓长江大保护"的各项决策部署下，围绕洪湖湿地保护与恢复工作，积极做好围网拆除、渔民上岸、生态修复"三篇文章"，实现了生态保护与经济社会发展的"双赢"。2018 年 4 月，习近平总书记视察湖北，亲临荆州，为长江经济带发展把脉问诊，为湖北、为荆州发展定位引航，具有里程碑意义。洪湖湿地保护工作注重把握三个关系，打赢三大战役，生态环境持续好转。

湖之精灵
fairies of the lake

灰翅浮鸥
Chlidonias hybrida

洪湖湿地　Honghu Lake wetland

把握整体推进和重点突破的关系，打赢围网拆除歼灭战。"绿水青山就是金山银山。"这句话在洪湖，体现在动真格、下真功的行动上。洪湖是湖北专业渔民最大的聚集区，20世纪80年代，全省大力发展水产养殖，洪湖围网面积最多的时候，高达25133公顷，占湖泊面积的71%，过度使用造成生态环境严重破坏。洪湖生态环境受到历届省委省政府的高度重视，为此，一场历时12年的拆除围网的战役吹响了号角。在这个过程中，政府全面作为，市县乡村四级联动，实施补偿措施，渔民深度配合。创造了"拆围洪湖速度"，也创造了单个湖泊一次性拆围面积全国第一。

把握生态环境保护和经济发展的关系，打赢渔民上岸攻坚战。拆围，只是洪湖生态环境保护的第一步，根治洪湖非法围网、非法捕捞的治本之策，还在于渔民上岸、转产转业。洪湖有拆围渔民3512户、12295人，其中易地扶贫搬迁（2016—2017）对象526户、1902人，渔民的命运从此翻开新篇章。洪湖市坚持"易地扶贫搬迁为基础、扶持就业创业为方向、健全社会保障为底线"的思路，解决了渔民的后顾之忧。通过保护区周边水产业升级，确保洪湖市作为全国淡水渔业第一县级市的地位。通过整治环境、建设环湖绿道、发展生态旅游，给地方经济注入新的活力，实现转型升级。

把握总体谋划和久久为功的关系，打赢生态修复持久战。洪湖，承载着长江四湖流域一万多平方千米范围内的上游来水，被誉为"江汉平原之肾"。作为长江大保护中的重要一环，保护洪湖责任重大。荆州市政府围绕"湖畅、水清、岸绿、景美"的目标，提出"流域治理、综合治理、工程治理、责任治理"的思路，坚持规划先行，确保洪湖治理取得成效；坚持综合治理，建立由17名湖长组成的四级湖长制，政府和市直部门组成的联席会议制度；实施全流域综合治理，全流域关闭污染严重的"十小"企业14家，封堵排污口54处，完成14家印染、造纸企业清洁化改造，70家畜禽规模养殖场和310家规模以下养殖专业户全部关停转迁。综合施策使洪湖生态治理取得成效。

水生植物管理　management of aquatic plant

碧水送北京　clean water to Beijing

五、丹江口水库——重要水源地与流域生态保护

1. 丹江口库区湿地及水资源

丹江口库区流域水系发达，河流众多，均属于长江的第一大支流汉江水系。河流均由北向南流向丹江口水库，主要河流有丹江、老鹳河、淇河、滔河等，大部分河道深，坡降大，水流湍急。

丹江口水库现状水质符合地表水水环境质量标准Ⅱ类标准，能满足各类功能用水的要求。丹江口库区及上游是南水北调中线工程水源区，是国家战略水资源保障区。

丹江口库区湿地是我国有代表性的湿地类型之一，是以河道型库塘生态系统为主要保护对象的内陆湿地和水域生态系统类型自然保护区。2000年被纳入《中国湿地保护行动计划》优先行动实施范围和重要湿地名录。

南水北调工程是我国优化水资源配置的一项重大战略举措。汉江流域丹江口库区作为南水北调中线工程核心水源区、国家战略水资源保护区、内河流域保护开发示范区、中西部联动发展试验区和长江流域绿色发展先行区，具有"融合两圈、连接一带、贯通南北、承东启西"的功能，在长江经济带高质量发展格局中具有重要战略地位和突出的带动作用。截至2021年5月，被誉为"中国水都"的丹江口水库已累计向北方京津冀豫调水 379 亿立方米，占南水北调总调水量的 87.9%，惠及 24 个大中城市约 7900 万人，实现了社会效益和经济效益双丰收。

2. 丹江口库区水源地保护与生态补偿

水源地保护与水资源保障。2018 年以来，国家"退耕还湿"政策在丹江口库区试点实施，坚持"全面保护，科学修复，合理利用，持续发展"的指导方针和"统筹规划，集中连片，优化布局，示范引领，逐步推进"的基本原则，抢抓时机开展栽植补造、生态补水、退耕还草，退耕还湿工程项目任务，933 公顷"退耕还湿"于 2020 年底全面完成，包括人工造林、人工种草、生态补水、封山育林等工程的实施，提升了库区森林覆盖率，进一步增强了湿地水源涵养和水土保持能力，优化了湿地生态系统的主导功能，改善水源区生态环境，为南水北调中线工程的生态安全提供有力保障，为南水北调中线工程向北方调取优质清洁的水资源做出了重要贡献。

"绿满荆楚"与"绿盾"行动。建设水源涵养林是提高水源保障能力、保护水

丹江口湿地　Danjiangkou wetland

南水北调中线丹江口水源区

Danjiangkou as a core water source area of the medium route project of the South-to-North Water Diversion

源区水质安全的重要途径。近年来，库区所在地的湖北十堰国土绿化以汉江和堵河及其支流两岸为重点，围绕荒山绿化、门户绿化、通道绿化扎实开展"水边、路边、村边"植树造林，全面完成了"绿满荆楚""精准灭荒"等重大工程，实施了长江、汉江两岸造林绿化专项战役，持续推进长江防护林、退耕还林工程，裸露山体修复治理及林相改造增绿添彩工作，为库区生态安全和水安全保障奠定了坚实基础；同时，开展"绿盾"专项行动，加强生态环境监管。

生态补偿机制与生态产业发展。把确保"一江净水永续北送"放在压倒性位置，积极探索实施库区生态补偿机制，推行重大工程项目过境保护区"事前审批、事中监管、事后监测"程序，落实生态补偿机制，确保了库区生态安全。充分发挥兴林富民社会责任，全力推进库区产业结构调整和效益提升，因地制宜培育主导产业，实现库区生态建设与产业发展双赢。木本油料、苗木花卉、林下经济、森林旅游等特色产业规模快速扩大，环库油橄榄产业快速发展，总规模突破5万亩，在省内外产生重要影响，北纬32°黄金产油带已现雏形。

六、襄阳汉江湿地保护联动，流域保护协同机制创先

汉江是湖北生态格局的重要支撑，是襄阳的母亲河，省域内蜿蜒流淌920千米。随着生态文明建设的持续推进，襄阳汉江走出一条整体性保护、系统性修复、综合性治理的绿色发展之路，呈现出清晰可见的生态之美、灵性之美、和谐之美，为大美湖北湿地增添亮丽的色彩。

襄阳汉江之美，美在一江清水。在大江大河中，汉江的水质是最好的。从监测数据看，整体水质稳定在Ⅱ类以上，一些指标接近或超过Ⅰ类水质，与丹江口水库中心区域的水质类似。开阔的水面，清澈的水体，加之周边200多处历史遗迹和众多传说，是汉江生态最大的特色，孕育出灵动之美。

襄阳汉江之美，美在水鸟逐浪。作为重要的候鸟迁飞通道和驿站，随着生态环境改善，汉江水鸟资源得到明显恢复。水鸟种类由五年前的78种增加到现在的132种，种群数量由过去9万余只增加到现在的16万余只，时常可见一群群几百只、上千只水鸟翻飞，野鸭闲游的胜景，展现和谐之美。

襄阳汉江之美，美在渚青岸绿，水波荡漾，水草菁菁，绿树成荫。汉江岸线自然延展披上绿装，岸边洲滩和江心洲岛形态各异，浑然天成，野趣盎然。500余

襄阳汉江国家湿地公园　Xiangyang Han River National Wetland Park

白骨顶
Fulica atra

白鹭
Egretta garzetta

黑鸢
Milvus migrans

湿地巡护　wetland patrolling

野外观鸟　birdwatching in the field

鸳鸯　*Aix galericulata*

种原生植物为湿地鸟类及其他动物构筑美丽家园。十年禁渔之策效果呈现，鱼群任意追逐遨游，构成汉江湿地原生态之美。

1. 创新汉江保护协作机制是汉江湿地保护之要

为更好保护湿地，从大流域着眼，汉江流域襄阳段建立起了湖北省首个湿地保护联盟——汉江（湖北襄阳段）湿地保护联盟。这是由湖北省湿地保护基金会发起，与汉江（湖北襄阳段）沿线六家国家湿地公园（襄阳汉江、谷城汉江、樊城长寿岛、宜城万洋洲、南漳清凉河、老河口西排子湖）共同协商、共同建立的。该联盟遵循"试点先行，逐步完善，协同发展"的路径，弘扬"共同使命、共同愿景、携手担当"的精神，为汉江湿地构建新保护格局，为新颁布湿地保护法更好地贯彻落地，为汉江流域襄阳区域联盟单位间合作、共享与创新融合，为更好地促进科学保护、信息共享、提档升级、联防共治，为保障汉江湿地生态功能发挥、生物多样性维护与生态安全更趋稳定，最终实现汉江流域生态文明建设纵深推行，实现人与自然和谐共生的汉江示范效应！

科学保护。建立湿地公园高质量保护与修复、改革创新等示范基地，推广典型，实现示范引路，整体推进。

信息共享。联盟成员单位通过科普宣教、科研监测、网络平台的共建，实现网络平台互联互通，信息、成果共享。

提档升级。推进汉江环境监测、资源调查，为湿地保护、科普宣教提供科学依据。

联防共治。联盟成员单位共同开展候鸟护飞行动、汉江水质保护行动。

2. 保持汉江水系连通性和流动性是汉江湿地健康之源

实施九水润城、古襄水修复、北河故道修复等重点工程，全域开展汉江水系连通工程，应对省内汉江6级水电开发，实行洪水与水量统一调度，保障汉江湿地水量充沛，健康顺畅，源远流长。

3. 开展十大战役等环境整治行动是汉江湿地安全之基

构建共抓汉江大保护工作格局，全面、深入、扎实开展沿江化工企业专项整治、农业面源污染整治、非法码头整治、非法采砂整治、沿江企业污水减排等十大标志性战役，为优美汉江湿地生态环境提供基础性保障。

4. 恢复和改善汉江原生态是汉江湿地发展之策

系统开展退田还湿、退养还湿、拆违还湿，实行人退湿进，实施十年禁渔恢复汉江食物链结构，重点实施鸟类栖息地恢复工程，重点推进原生植被恢复，有效提升汉江湿地生态系统质量和稳定性。

5. 打造人民群众共享的绿色空间是汉江湿地惠民之路

襄阳汉江六家国家湿地公园不断加大建设力度，打造优美生态环境，提供优良生态产品和优质生态服务，把高品质的自然教育、研学基地与生态休闲有机结合，使之成为市民放松身心，拥抱大自然的理想之所。

七、沉湖国际重要湿地——智慧湿地提升科学管理水平

1. 沉湖国际重要湿地及自然保护区

沉湖湿地于1994年1月建立区级自然保护区，次年晋升为市级，2006年8月晋升为武汉市第一个省级湿地自然保护区。2013年10月被确定为国际重要湿地。沉湖国际重要湿地（湖北沉湖湿地省级自然保护区）位于武汉蔡甸区西南部，是以典型湿地生态系统和珍稀水禽为主要保护对象的内陆湿地和水域生态系统类

型自然保护区。主要由沉湖、张家大湖、王家涉湖和杜家台分蓄洪区部分区域组成，是江汉平原典型的淡水湖泊和沼泽湿地。

保护区内野生动植物资源丰富，有维管束植物330种，其中国家二级重点保护植物4种；底栖动物73种，鱼类58种，两栖类10种，爬行类28种，哺乳动物27种；鸟类资源277种，其中国家一级保护鸟类14种，国家二级保护鸟类50种，省级重点鸟类47种，每年在此停歇、繁殖或越冬的候鸟逾10万只，有18种鸟类分布数量超过全球种群总量的1%，被专家誉为"湿地水禽遗传基因保存库"。

2. 沉湖生物多样性监控与智慧湿地建设

（1）沉湖生物多样性监控。沉湖是重要的鸟类栖息地，生物多样性比较丰富。保护区积极与科研单位及武汉观鸟会等环保组织合作，开展越冬季鸟类专项监测及常规鸟类监测、湿地生物多样性监测与评估、湿地生态环境地质调查监测、湿地野生动物疫源疫病监测防控等科研监测项目，为沉湖湿地保护管理提供科学决策依据。

（2）沉湖智慧湿地平台建设。沉湖生物多样性监测与智慧湿地平台建设是以生态学、保护生物学基本原理为依据，集合了当前最先进的人工智能、物联网、大数据、云计算等技术打造的系统解决方案，实现各种生态数据实时分析和生态预警，是武汉市首个智慧湿地平台。

沉湖　Chenhu Lake

小天鹅
Cygnus columbianus

白鹤
Grus leucogeranus

灰鹤
Grus grus

生物多样性感知网络。设计建设了较为完善的感知网络，涵盖了天空地不同尺度、水土气生不同生态要素的监测需求。

湿地生态大脑。包括沉湖湿地实景动态模拟、野生动物视频识别、野生动物声纹识别、遥感影像智能解译软件、物种卫星追踪、监测信息分析预警、管理活动指挥等各模块。

保护区信息中心。建设了会议室和展示大屏，组建了保护区自有的信息机房用于实时处理感知数据，实现沉湖湿地生态系统智能化管理。

八、精准布局创新亮点，三位一体聚合效益——天门张家湖湿地

张家湖国家湿地公园自然环境优美、生态区位重要、文化底蕴深厚、湿地资源极其丰富。区域内河湖纵横交错、平岗跌宕起伏、物产丰富，素有"鱼米之乡，圣水宝地"的美誉。张家湖是国家野莲的保护与持续利用推广点，其"水八仙"，即野菱、慈姑（*Sagittaria trifolia*）、荸荠、野莲、芡、水芋（*Calla palustris*）、水芹（*Oenanthe javanica* DC）、茭白生态种植基地产品闻名遐迩，其还是天门市三大水

张家湖国家湿地公园　Zhangjia Lake National Wetland Park

产种质资源保护区之一——四大家鱼保护区，出产的青、草、鲢、鳙（*Aristichthys nobilis*）、乌鳢（*Channa argus*）、中华绒螯蟹（*Eriocheir sinensis*）、鲤（*Cyprinus carpio*）、鲫（*Carassius auratus*）等8种鱼蟹被农业部认定为水产绿色食品。

天门市委、市政府将张家湖国家湿地公园及生态产业开发的基础设施建设，作为全市生态文明重点建设项目在持续推进中。以保障张家湖生态功能良性提升为基础，遵照国家公园"全面保护、科学修复、合理利用、持续发展"的建设方针，天门市开展了"三位一体"项目整体建设的精准布局：突出文化休闲体验特色，坚持以本土资源和创新力为抓手，整合内外部资源链接、优势互补，着力湿地生态系统科学保护恢复管理、科研监测与科普宣教、湿地产业基础设施三大核心建设，打造我国平原湿地生态系统重要示范工程、湿地生态保护教育重要示范基地、以张家湖国家湿地公园为核心的湿地产业基地。

1. 创新管理体系，构建全面保护支撑力

构建全市组织管理支撑体系，健全管理机制保高质量发展。为加强湿地公园生境保护，天门市委、市政府成立了国家湿地公园保护工作领导小组，由市委书记任政委、市长任组长，下设畜禽养殖场取缔、农村生活垃圾综合整治、农村面

黑枕黄鹂　*Oriolus chinensis*

普通翠鸟　*Alcedo atthis*

牛背鹭　*Bubulcus ibis*

白鹭　*Egretta garzetta*

野莲　*Nelumbo nucifera*

源污染整治、"厕所革命"、河湖"清四乱"和河湖长巡查等工作专班，分工协作、统一行动，严格执法、常态管理，有力推进了国家湿地公园保护建设管理工作生态化、严细化、高效化。同时，成立了湖北天门张家湖国家湿地公园管理局和天门张家湖生态资源保护开发有限公司，承担湿地公园保护、建设、营运等职能。

2.创本土特色科普，引公众意识入胜

建立人文科普宣教体系与阵地。着眼公园宣教阵地的氛围建设与集中宣教效应，公园在一期建设中就增设了湿地科普宣教馆、湿地自然学校、宣教长廊、标识系统、湿地广场等宣教设施。利用主题鲜明的宣教长廊，湿地标识标牌、旅游交通导识牌，以寓教于乐的方式让湿地访客、市民了解张家湖湿地人文特色、湿地百科及本土动植物资源等湿地知识。

让人文科普活动走入公众。坚持利用"世界湿地日""爱鸟周"等时间节点，面向公众开展主题鲜明、内容丰富的科普公益活动，促进湿地科普知识进机关、进学校、进企业、进社区、进家庭，由此普及了湿地知识，增强了公众湿地保护意

湿地科普　science education of wetlands

湖北天门张家湖国家湿地公园科普宣教中心
science and education center in Hubei Tianmen Zhangjia Lake National Wetland Park

张家湖湿地　Zhangjia Lake wetland

识,弘扬了生态文化,促进了人与自然和谐共生。张家湖国家湿地公园还被湖北省科协命名为湖北省科普教育基地。

打造生态文明教育基地。充分利用湿地宣教中心、湿地课堂、湿地植物认知园、观鸟塔等宣教设施,面向公众开展湿地生态研学、自然教育活动,使其成为天门市新晋"国字号"景点,市民生态文明教育基地。

引导社会力量组织开展不同形式人文科普活动。组织湿地公园周边9个村、社区居民成立志愿者工作队,参与湿地保护,携手共建美丽家园;充分利用市摄影家协会、老年书画家协会、诗词楹联学会等社会力量的专业优势,为他们提供创作基地平台,广泛培植科普队伍。

融合媒体,形成广泛传播影响。利用自媒体平台,及时向公众发布湿地公园建设成果、美丽风光、保护措施和科普知识。充分运用网络、报纸、电视等媒体开展科普宣传,广泛宣传湿地保护知识,大力倡导生态文明理念。

3. 创湿地经济品牌,建生态产业基地

探索共建共管模式。充分调动周边群众参与带动周边产业发展,深入社区开展共建活动,先后召开座谈会20余次,与周边村签订了共建共管协议,建立了日常沟通机制,定期组织村民参观考察,增强全民湿地保护意识。

打造湿地产业基地。在周边9个村着力打造"水八仙"特色产业基地,构建"基地+农户+公司(合作社)+电商平台"运营模式,全力打造"水八仙"电商产业园,助推农村经济持续健康发展。

九、传承发扬生态圩田智慧,再现孝感朱湖梦里水乡

湖北孝感朱湖国家湿地公园位于孝感市孝南区,总面积5156公顷,以平原湖区为主,主要由河流湿地、沼泽湿地、人工湿地三大类组成,涵盖永久性河流、草本沼泽、库塘、输水河、稻田五个湿地型。2018年12月获国家林业和草原局批准为国家湿地公园;2018年12月被评为国家3A级旅游景区,同月,被省科协命名为湖北省科普教育基地;2019年获批"湖北省重要湿地";2021年10月被列入全省近零碳排放示范工程试点,是湖北省摄影创作基地,省内多家重要旅游公司定点生态旅游目的地。

1. 朱湖特色的退田还湖与生态修复之路

孝感朱湖公园退耕还湿后,展现出"人退湖进,人护湖盛"的景象。

首先,实施朱湖流域层面的环境综合治理措施,先后实施府沦河抗洪排涝、灾

朱湖湿地　Zhuhu Lake wetland

后水利设施重建、府河堤加高培厚、府澴河水环境综合治理、生态保护区拆迁还建、渠道清淤、内渠渠网连通等湿地保护与生态恢复治理工程。其后，通过退耕还湿、退田还湖，走出有朱湖特色的生态修复之路。建立向阳垸多功能圩田，培育500多块规模不等和形态各异的湿地食用水生植物田、湿地观赏植物田、湿地药用植物田、湿地编织用材料植物田、湿地种子植物田等应用类型，以丰富生物多样性，促进水土涵养、水质净化、生物保育、调蓄洪水、调节小气候功能的提升。建立基塘、多塘系统，在公园周边建设乔灌结合，林带、林园组合的林泽复合湿地工程，并根据地形地貌，在原有农田系统基础上，因湖就水种植大量本土植物，丰富生物多样性，充分利用空间，吸引更多的生物栖息。修复生态链条，在湿地公园保育区和恢复重建区，补植芦苇、菖蒲（*Acorus calamus L*）、水烛（*Typha angustifolia L*）、水葱（*Scirpus validus Vahl*）、莲等30多个品种的本土野生植物，建立微小湿地群500多个共780亩，充分发挥了涵养生态、净化水质、孕育物种、营造生境、美化环境五大功能，形成大湿地套小湿地、小生境提升大环境的良好状态，营造别具一格的自然和谐、互利共生的生态环境，取得了生态保护与恢复的独特效果，芡实、水葱、莼菜等10多种一度在朱湖区域销声匿迹的野生植物，青头潜鸭、小天鹅、白额雁、苍鹭等20多种数十年不见的野生鸟类再现河湖渠汊，"才闻鸟鸣在东陌，又见野莲卧菱湖"的生态美景，生机盎然，动人心弦。

2021年，朱湖国家湿地公园被列入全省近零碳排放示范工程试点后，围绕湿地公园，形成了由3万亩国家地理标志保护产品"朱湖糯米"种植基地、2万亩生态水产健康养殖示范场、2万亩优质高效花卉苗木产业基地组成的生态共同体。

2. 朱湖湿地圩田与湖乡生态产业

圩田是我国古代劳动人民利用濒河滩地、湖泊淤地过程中发展起来的一种农田。人们在低洼地区四周筑堤，并在堤上设置涵闸，平时闭闸御水，旱时开闸放水入田，这样无论旱涝都可以保证收成。圩田，曾经在朱湖发展中担任过重要角色，呈现出"畦畛相望""阡陌如秀"的景象。如何传承好圩田文化，让今天的新型生态圩田在朱湖湿地保护和生态农业发展中担当起应有的历史使命？

放大湿地特色优势。恢复传统圩田，创建生态圩田，充分发挥生态圩田具有的生态产品产出、环境优化、水质净化、生物保育、调蓄雨洪等多种功能，打造富有朱湖特色的湿地生态保护模式。

圩田春色
a spring sight of polder

多功能生态圩田
multi-functional ecological polder

湿地生态旅游
wetland ecotourism

放大湿地合理利用功能。在生态圩田区域规划设计了500多块大小不一的湿地食用水生植物田、湿地观赏植物田、湿地药用植物田、湿地编织用材料植物田等应用类型，以丰富生物多样性、水土涵养、水质净化、调节小气候，实现人与湿地的协同共生。

放大湿地人文景观。把握古云梦泽历史遗韵，将原有矩形农田改造恢复为古代圩田形态，为水生植物、底栖生物、鱼类和水鸟提供了理想的栖息地，同时也成为观赏性较强的湿地人文景观。近年来，先后吸引110多批次作家、诗人、文学爱好者，以及新闻记者、新媒体创作人员到朱湖湿地观赏采风，在国家及省区市各级媒体、报刊发表各种体裁的湿地作品500多种，摄影作品1000多幅。

放大湿地自然景观。通过发展和精心培育湿地生态圩田，拓宽了建设视野，圩田植物达30多种20多万株。目前，湿地"五区"和5万亩稻田共同组成一个庞大的生态循环系统，湿地生态环境基本恢复到20世纪70年代初期水平。

针对朱湖发展潜力和区位优势，将湿地经济发展的方向定位在培育名优产品、开拓特色经济、兴办产业基地、保护生态环境、推进农旅融合上，突出重点调整产业结构，优化区域经济布局，形成四季林木葱郁、糯稻飘香、水清鱼跃、观光休闲的农旅融合产业园。"朱湖糯米"是朱湖名特优产品，也是朱湖湿地生态系统中十分重要的一环，朱湖及周边拥有5万亩稻田，常年气候湿润，水美稻香，是白鹭、野鸭、大雁等鸟类出没的胜地，以其高品质的湿地功能和高效益的经济功能成为湿地公园引人陶醉的田园观光景致。湿地公园是一块孕育和传承湖乡文化的沃土，这里是孝感米酒、麻糖、糍粑、豆皮等农家传统食品制作的发源地之一，在建设湿地公园过程中，大力支持民营业主投资3000多万元创办了孝感市天龙酒业有限公司，专业从事"孝感米酒"系列产品的研发和生产，利用湿地优质糯米和湿地优质水源先后开发出16种软包装、休闲式、新口味的"孝感米酒""珍珠糯酒"，实现了"孝感米酒"从这里走向全国，走向美国、新加坡、哈萨克斯坦等世界多国。

十、远安沮河——全域湿地保护与乡村振兴融合管理

湖北远安沮河国家湿地公园遵循"全面保护、科学修复、合理利用、持续发展"的方针，积极探索"绿水青山就是金山银山"的转化路径，提高湿地生物多

样性及复合生态系统的稳定性，提升全社会生态环境保护意识，为建设生态文明和美丽远安提供重要保障。

1. 远安沮河湿地与资源特色

湖北远安沮河国家湿地公园位于沮河中游、县域中南部，紧靠远安县城，包括区域内沮河、鸣凤河、九子溪河道及两岸山体、林地和部分农田，水岸线全长59.05千米，是沮河流域湿地生态系统的重要组成部分，在鄂西地区乃至全国都具有一定的典型性和代表性。2019年和2020年，远安沮河湿地先后列入《湖北省第一批省级重要湿地名单》《2020年国家重要湿地名录》。

远安沮河湿地物种资源丰富，有各类植物448种，其中国家重点保护植物8种；有脊椎动物257种，鸟类达到173种，其中国家重点保护动物22种，国家一级重点保护动物有黑鹳、中华秋沙鸭、东方白鹳和白冠长尾雉（*Syrmaticus reevesii*），鱼类43种，其中有13种中国特有鱼类，另有省级重点保护动物44种，列入国家保护"三有"动物166种，湿地内分布的8种两栖动物均被列入"三有"动物名录。

远安沮河两岸　both sides of Juhe River in Yuan'an

中华秋沙鸭　*Mergus squamatus*

2. 全域湿地保护与管理创新

一盘棋管全域的湿地保护。远安县建立谋全局、管全域的统筹协调机制，构建湿地公园建设和湿地保护管理全县一盘棋的工作格局，重视生态治理和湿地保护修复，树牢全域生态理念，将湿地保护从局部试点向全域拓展，把湿地保护与全域旅游、乡村振兴、城乡建设和水利、交通基础设施等各个领域深度融合；明确提出"生态优先、全面保护，尊重自然、体现乡愁，综合治理、低碳治污，突出功能、科学修复"的湿地保护与建设思路。

全方位织密湿地保护网络。将湿地公园保育区和恢复重建区纳入生态红线管控，全方位落实管护力量，构建"管护员+河长+志愿者"的湿地公园巡查管护网；采取湿地公园管理处和远安县林业局"全员下沉一线"的方式开展巡护监督；强力推进联合监管，联合县生态环境分局、水利局、农业农村局等部门，全面开展"清四乱""迎春行动""清流行动"等专项行动，实施沮河流域生态环境整治；设立琵琶洲、陈家棚野生动物栖息地保护点，建立湿地公园视频监控平台，设置监控点，规范完善管护设施，基本实现公园河道在线监控全覆盖，公园保护管理水平持续提升。

3. 推动小微湿地保护，融合乡村振兴和全域旅游

牢固树立人与自然和谐共生的理念，把生态文明建设放在更加突出的位置，务实做好生态修复、环境保护、绿色发展"三篇文章"，着力让优良生态环境成为远安高质量发展的重要支撑点和增长点。坚持节约优先、保护优先、自然恢复为主的方针，深刻把握湿地被称为"地球之肾"所表达的重要内涵，将湿地公园设计理念提升到乡村振兴战略实施根本遵循的高度，推动小微湿地建设工作全面融入绿色城镇建设和全域旅游发展，以提升生态功能和生态价值为导向，合理利用乡土自然资源，科学推进小微湿地建设和保护，着力守护生态本底。

开展全县小微湿地保护模式。通过自然与人工相结合的方式进行生态修复，保护湿地资源，恢复生态功能。先后建成旧县镇鹿苑河董家段、洋坪镇五里河芭芒

远安七彩马渡河　colorful Madu River in Yuan'an

店、鸣凤镇西湖塘堰、马道堰、鸣凤河口、沮河三桥北门段退耕还湿等小微湿地。

推动特色小微湿地利用模式。保护湿地资源与促进农业开发相结合，建设生产型小微湿地。鼓励引导农民及农村经营主体发展涉水特色种植、养殖，推动小微湿地建设与保护，促进休闲农业和乡村旅游发展，实现经济价值和生态价值双提升。先后建成金家湾乡村小微湿地、桃花岛生态公园等生态湿地景区，以及鸣凤镇九子溪小区、双泉小区、旧县镇洪府庄园等庭院小微湿地。

探索湿地宣教和社区共建模式。公园紧密聚焦乡村振兴这一战略布局，努力探索社区共建模式，公园先后建成湿地自然学校、湿地自然图书馆、湿地植物宣教长廊，并将生态文明理念融入群众日常生活之中，让群众在休闲健身中体会到浓郁的生态文化氛围。湿

金家湾小微湿地
Small wetlands in Jinjiawan

十一、践行两山论——探索徐家河三生融合高质量绿色发展之路

1. 徐家河湿地功能特色与水源地保护

广水徐家河国家湿地公园依托徐家河水库而建,属涢水水系,上游主要河流有徐家河、龙泉河等四大支流,均发源于桐柏山南麓,汇入徐家河水库。徐家河水库是湖北省第三大水库,原水源补给状况以大气降水和地表径流综合补给为主,现辅以鄂北水资源配置工程丹江口补水。水库总库容 7.78 亿立方米,兴利库容 2.99 亿立方米,多年平均入库流量 6.7 米^3/秒,水库流域内多低山丘陵,岛屿、库汊多,是典型的丘陵分汊型水库。徐家河水库号称"鄂北明珠",集水源、生态、防洪、蓄洪、旅游等多种生态功能于一身。

徐家河国家湿地公园具有生物多样性、库塘湿地景观与人居环境协同共生的复合湿地生态系统型的湿地公园,其多重湿地复合形成特有湿地生态系统。徐家河水库水域面积广阔,湿地野生动植物资源丰富,水库湿地特征显著,湿地形态

徐家河水库　Xujiahe reservoir

自然，景观秀丽，观赏性强，是长江中下游地区湿地的重要组成部分，具有极强的代表性。

徐家河水库是重要的水源地，承担着区域重要的饮用水源和灌溉水源功能。广水市政府制定了《广水市徐家河水库综合整治实施方案》，开展徐家河湿地保护与综合整治行动，对徐家河湿地公园内及周边持续开展了一系列环境整治和生态治理工作，取得了较好的成果，显著改善了徐家河水生态和水环境。一是取消网箱养鱼，拆除、取缔徐家河水库网箱、拦网，彻底清理水面养殖。二是整顿畜禽养殖，查处各类养殖场（养殖户），控制污染源。三是禁止河砂开采，在全市发布关于打击徐家河库区非法采砂的通告，严格落实河长制及部门职责，严厉打击库区非法采砂行为。四是整治违章建筑，拆除违建别墅等违章建筑，复垦植绿，排除污染隐患，保持岸线的生态完整。五是启动徐家河水库禁捕工作。六是增殖放流，投放鲢、鳙、草、鳊等鱼苗，丰富了徐家河生物资源。七是开展入库生活污水治理工程。八是拆除非法矮围，极大地保证了水系畅通、水域联通，防止了养殖污染，改善了湿地生态环境。

目前，徐家河灌区涉及广水、安陆、云梦、孝南、孝昌五个县市，农田灌溉面积3.8万公顷，是一座以防洪、灌溉为主，供水、航运、养殖等综合利用的大（2）型水库，同时也是广水、安陆、云梦、孝昌、曾都等县市区100多万人口饮用水和生活用水的水源地。

2. 三生融合发展理念下社区共建模式

打造湿地周边美丽乡村，突出创建特色。将湿地公园建设与美丽乡村建设有机衔接，稳妥推进湿地公园与社区共建，进一步打造了樱花狮子、花韵泉水、人文肖店等一批湿地与社区共建点，突出创建特色，让湿地生态保护与建设同周边居民生产生活相得益彰。

（1）马坪镇狮子岗村共建站点。狮子岗村位于马坪镇东部，徐家河国家湿地公园西岸。狮子岗村先后获得国家森林乡村、省级绿色示范村、省级移民美丽家园和省级卫生村等荣誉称号。村内还建成了占地60余公顷的"樱花园"。狮子岗村现已具有良好的生态环境和生态旅游价值，湿地公园与该村共建管护站点，将进一步加强徐家河湿地保护力量，提升湿地生态环境。

（2）关庙镇梅庙村共建站点。梅庙村位于关庙镇东南部，徐家河国家湿地公园

梅庙生态旅游村
ecotourism village of Meimiao

泉水村
Quanshui village

徐家河湿地桃花岛
Taohua island in wetlands in Xujiahe reservoir

北岸。梅庙村先后获得"全国乡村治理示范村""国家森林乡村""湖北省生态旅游村"等荣誉称号。村内建成了占地1000余亩的"格桑花园",500余亩的"月季园"以及其他苗木园地1000亩。梅庙村具有优美的生态环境和管护队伍的基础,与该村共建管护站点,将以最少的投入,提升湿地公园的管护能力,扩大湿地科普宣教的影响力。

(3)长岭镇泉水村。泉水村位于广水市长岭镇东北部,徐家河国家湿地公园东岸。泉水村已获得AAA级旅游景区,先后荣获"湖北省绿色示范乡村","随州市十大旅游名村",还荣获了"湖北省十大荆楚最美乡村提名奖"等荣誉称号。在村内积极发展花卉苗木种植和生态养殖产业。泉水村围绕"美丽乡村花韵泉水"打造出了良好的生态旅游景点,与该村共建管护站点,将进一步提升徐家河国家湿地公园的生态旅游形象,为后期的旅游发展奠定基础。

(4)长岭镇新庵村(明月湾度假村共建站点)。长岭镇新庵村(明月湾度假村)位于广水市长岭镇中部,徐家河国家湿地公园南岸。明月湾度假村已获得AAA级旅游景区,是广水市作家协会创作基地、广水市诗词楹联散曲创作基地、随州市研学旅行基地。获得湖北省文化和旅游厅授予的国家五星级农家乐、湖北省休闲农业示范点等荣誉称号。明月湾度假村由广水市徐家河水浒文化旅游有限责任公司于2004年投资兴建,是以诗仙李白文化为核心,聚休闲观光和农业科普、研学为一体的乡村生态旅游目的地。明月湾度假村现为徐家河湿地周边最具影响力的生态旅游景点,有良好的生态环境和管护基础,与该村共建管护站点,将进一步加强湿地生态保护力度与宣传力度。

3. 高质量迈向湿地全面保护和乡村振兴

广水市探索实践"两山论",创建高质量融合发展示范区。广水市提出"建设月光海高质量融合发展示范区"的目标,结合创建徐家河国家湿地公园建设和管理,制定行动计划,加快建成"两山论"实践区、乡村振兴示范区、生态保护标杆区、产业融合样板区、基层治理模范区的基本要求,最终实现区域内人民群众共同富裕的重要保障。

规划引领,科学策划。以立足保护首位、生态优先、规划引领、全域联动、分步实施、重点打造的原则,牢固树立"两山"理念,确立了以国家旅游度假区、国家湿地公园为切入点,建设"六区一园"的融合发展之路。通过引入高水平策划团队，制定了示范区创建行动计划，确定了实施方案，启动了规划策划，筹建了起步项目，稳步推进湿地公园生态保护与建设，为高质量融合发展开好头、起好步。

梳理政策、盘点资源。一是政策要求，包括湿地保护法的规定和要求，饮用水源地保护要求和范围，湿地公园范围和功能分区要求，水库管理范围及要求，银鱼种质资源保护地范围及要求等。二是文化底蕴，主要是李白寿山文化、詹王饮食文化、水库渔民文化、云台山红色文化。三是存量资源，对交通培训中心、财校、人事厅、疗养院、六角岛、桃花岛、龙凤岛7处闲置可用资源，进行了全面细致地盘点，做到一处一本台账。四是特色产业，示范区重点培植的产业是茶叶、油茶、蔬菜、香稻、油菜、华鼠尾草中药材，以及花卉苗木等。

加强示范区自然资源保护和管控。制定《关于加强示范区自然资源保护利用、规划建设管控工作的通知》，以指挥部的名义下发相关成员单位和五镇办。

十二、汉川汈汊湖——退垸还湖重现自然生态美景

1. 汈汊湖演变与湿地现状

汈汊湖位于汉江下游、江汉平原北缘、湖北省汉川市北部，它是由东西南北四条干渠环抱所形成的人工封闭型湖泊，外部受四条干渠的严格控制，呈现为东西长16.1千米，南北宽5.5千米的长方形。汈汊湖原本是一个泛洪区，是一个水来一片汪洋，水退一片凄凉的湖泊。在20世纪50年代之前，这里大小湖泊组成湖泊群，原天门河、溾水、四龙河、大富水、涢水等多条中小河流经汈汊湖注入汉江。

1969—1972年，汈汊湖被围垦成全国最大的内陆封闭式淡水湖泊，其主要功能是排洪削峰，保障流域人民财产安全。当时，汈汊湖遵循"调蓄为主结合养殖"的开发方式，近百平方千米的汈汊湖被围成了星罗棋布大小不一的精养鱼池，湖面加快萎缩。

2014年，国家林业局批准汈汊湖作为国家湿地公园建设试点；2017年，汈汊湖被列为湖北省河湖长制单位；2018年，汈汊湖被列为长江大保护重点整治项目；目前，旨在恢复和保护汈汊湖原有湖泊湿地生态系统，合理利用湿地资源，汈汊湖迎来了"六湖飞歌"（退垸还湖、法治安湖、生态美湖、产业兴湖、特色立湖、创新强湖）的历史机遇。

退垸后的汈汊湖湿地

Diaochahu Lake wetland after returning fields with protective embankments for levees

2. 汈汊湖退垸（田、渔）还湖与生态修复

2019年3月，湖北省人民政府批复10个跨界湖泊的保护规划（以下简称《规划》），汉川汈汊湖被纳入其中。该《规划》的实施将有效保障湖泊形态稳定，水面（容）积不减少，开发利用得到有效控制，湖泊生态实现良性循环，湖泊综合功能得到充分发挥，实现湖泊生态保护和资源可持续利用。汉川汈汊湖"三退"暨国家湿地公园建设是汉川市委市政府向生态要绿色发展，造福子孙后代的战略决策。

开展退垸（渔、田）还湿工程。汈汊湖退垸（渔、田）还湖（三退）工程实施范围为蓄洪区（养殖区）共48.7平方千米，涉及6家市直单位、汈汊湖养殖场7个行政村、9个国营生产队，约2189户，总共发放补偿资金11.78亿元。此外，新建还建小区安置上岸渔民，发放养老金解决年岁较大渔民养老问题，开展创业培训、提供就业岗位帮扶上岸渔民再就业。

开展生态治理与生态修复。生态治理主要包含水系连通、清淤疏浚、边堤加高培厚。水系连通主要是间歇性拆除了恢复重建区内原养殖池塘的堤埂，使得湖内水体相互连通，共拆除堤埂长为18.5千米，土方347.1万立方米。生态修复工程包含滨湖陆生植被修复、湖泊水生植被修复，植物过滤带、浅滩与鸟岛、栖息地恢复等工程。

3. 汈汊湖生态修复与治理成效

汈汊湖累计拆埂还湖 4000 多公顷，实现了调蓄区 48.7 平方千米鱼池全面"三退"，投放鱼苗 100 万尾，种植狭叶香蒲等挺水植物 66 万平方米；湖面比"三退"前扩大约 20 倍，湖泊生态和相关功能大幅改善，汈汊湖调蓄洪能力极大增强，达到 1.2 亿立方米。

目前，通过开展退垸（田、池）还湿、拆除围网、水系连通、清淤疏浚、水生植被恢复等系列湿地保护与恢复工作，汈汊湖生态环境持续改善，为野生动植物繁衍生息创造了良好条件。汈汊湖水质已从Ⅴ类上升到Ⅲ类标准；区域内野生动植物种类、数量都有明显增加，有新增维管束植物 67 种，鸟类 9 种，小天鹅、白琵鹭等国家保护动物种群数量增加十余倍。

为全力推进汈汊湖"三退"工作，政府积极探索"湿地+公园""生态+旅游""创业+就业"的高质量绿色发展路径，充分依托汈汊湖流域的自然资源特性，借力全域旅游国家战略，做好汈汊湖水湖林田草、鱼虾兽虫鸟湿地保护文章；结合生态文化、水利文化、渔业文化、历史文化、科普教育，打造独树一帜的全国人工内陆"第一湖"；同时，确定安置补偿方案，解决湖民的生计问题。

小天鹅 *Cygnus columbianus*

十三、绿色赤壁——湿地保护与区域发展新实践

1. 赤壁市域典型湖泊湿地

位于长江中游南岸的赤壁,有着深厚的历史文化积淀和丰富的地理自然资源。以陆水湖、黄盖湖和西凉湖为代表的湿地,构建了极其优良而又独具特色的湿地生态系统。

(1)陆水湖,水域面积57平方千米,蓄水量7.2亿立方米。湖中800多个岛屿星罗棋布。陆水湖主坝位于赤壁城区陆水上游,属三峡工程试验坝,八号副坝是亚洲最长的人工黏土坝,主坝下的陆水河在赤壁的青山和田野蜿蜒40千米后从三国赤壁之战遗址所在地赤壁镇注入长江。陆水湖水质优良,水产丰富,集蓄水、调洪、发电、养殖于一体;湖区动植物种类繁多,山清水秀,风景宜人。

(2)黄盖湖,位于湘鄂两省交界处,北接长江,西邻洪湖,为湘鄂两省共有的天然湖泊。历史上的黄盖湖曾是古云梦泽的一个不见经传的湖泊,东汉末年的赤壁大战,不仅让赤壁石矶名扬四方,也让这个千年古云梦泽上的水洼有了一个自己的名字——黄盖湖。黄盖湖湿地独特的地理位置和生态环境,孕育了丰富的湿地动植物资源,而且有许多珍稀濒危物种,并且呈现珍稀物种较高,种群数量大的特

赤壁陆水湖国家湿地公园　Chibi Lushui Lake National Wetland Park

黄盖湖　Huanggai Lake

点，区内辽阔的水域和宽阔的湿地是鸟类越冬迁徙的重要中转站和栖息地。同时，黄盖湖是一个通江湖，全年有9个月与长江连通，物种资源可以自由进出和交流，形成了丰富的湿地生物资源，为物种的多样性提供了便利条件。

（3）西凉湖，位于赤壁市东北角，地势低平，水网稠密，河汊相连，有82平方千米广阔水域（赤壁市沿湖乡镇约占西凉湖水域总面积的60%，嘉鱼县和咸安区沿湖乡镇约占西凉湖水域总面积的40%）。西凉湖的莲藕、芡、蒿笋、菱、莼菜四季葳蕤，青鱼、鲤、草鱼、鲫、鳊、鳜、鱤、乌鳢、鲨、红尾鱼和黄鳝、龟、鳖、蟹、虾等诸多淡水鱼类畅泳其间，也是野鸭、白鹭等一众水鸟筑巢做窝繁衍生息的场所。

2. 赤壁湿地保护管理发展实践

作为湖北省首批建设的国家湿地公园，陆水湖国家湿地公园从成立的那一天起，就承担着绿色发展的历史使命。2017年1月，根据市委市政府工作部署，明

赤壁陆水湖国家湿地公园　　Chibi Lushui Lake National Wetland Park

确了陆水湖国家湿地公园管理处的职责分工，承担包括陆水湖在内的全市湿地保护与管理工作。

作为自然保护地的一种形式，国家湿地公园的建设都是按照一地一策的原则进行的。从某种意义上讲，赤壁陆水湖国家湿地公园的管理模式，开启了湿地保护发展的新模式、新途径。

实践已经证明，赤壁湿地保护发展的新实践，避免机构的重复建设，集约使用财政资金，让专业人做专业事，不仅让赤壁湿地资源得到了有效的保护，更可贵的是，这种新实践有效对接了国家正在推进的以国家公园为主的自然保护地整合优化改革，呈现出极强的生态保护成效和区域绿色发展活力。

2016年以来，根据国家对长江流域"共抓大保护、不搞大开发"和"十年禁渔"的指示精神，在西凉湖、黄盖湖等湖泊，赤壁市启动了以拆除湖域全部"楠竹（竹桩）、网片（网箱）、地笼、迷魂阵"和"渔民上岸"为主要内容的"禁渔禁捕"集中攻坚行动。随着一根根楠竹竹桩的拔除和"十年禁渔"的全面实施,"落霞与鹭鸟齐飞,碧水共蓝天一色"的西凉湖和黄盖湖,用水的澄澈和清纯,风姿绰约地呈现出山环水绕、百转千回的旖旎风景。

3. 赤壁湿地生态保护成效

黄盖湖优越的地理环境、湿润的气候孕育了丰富的生物多样性。在黄盖湖庞大的生物量记载中,鸟类几乎占据了头版,以冬候鸟为例,每年到黄盖湖地区越冬的水鸟数量达到 4 万～5 万只。冬春交替之际,正是冬候鸟的高峰之时,无论是宽广无垠的水面,还是水泽滩涂间,我们都能找到鸟儿的身影。仅 2021 年的秋冬季科考记录中,已调查记录鸟类数量 67757 只,隶属于 15 目 42 科 128 种,其中,水鸟 7 目 11 科 41 种。调查记录到国家级重点保护鸟类 7 种,其中有国家一级保护动物白鹤 1 种,国家二级保护动物小天鹅、黑鸢（*Milvus migrans*）、白尾鹞（*Circus cyaneus*）、红隼（*Falco tinnunculus*）、游隼（*Falco peregrinus*）、小鸦鹃（*Centropus bengalensis*）共 6 种。黄盖湖丰富的水鸟资源和高度的物种稀有性保护价值,已经引起全社会的高度关注。

东港湖湿地　Dongganɡ Lake wetland

十四、城市湖泊生态修复，打造湖北湖泊治理样板

根据 2022 年 7 月湖北省水利厅发布的《2021 年湖北省湖泊保护与管理白皮书》，列入省级湖泊保护名录的 755 个湖泊中，完成生态修复规划的湖泊有 293 个，正在实施的湖泊生态修复项目有 120 个（其中清淤及综合治理项目 86 个）。2021 年全省主要湖泊总体水质为轻度污染，主要污染指标为总磷和化学需氧量。24 个省控湖泊的 29 个水域中，水质为Ⅱ类、Ⅲ类的水域占 31.0%，水质为Ⅳ类、Ⅴ类的水域占 69.0%，无劣Ⅴ类水域。29 个湖泊水域中，2 个水域营养状态级别为中营养，27 个水域为轻度富营养。与 2020 年同期相比，可比的 18 个湖泊水域中，水质同比好转水域的有 2 个，保持稳定的有 10 个，有所下降的有 6 个。湖北省湖泊保护与管理工作任重道远，生态修复工作大有可为。

在武汉的自然环境中，湖泊是最有特质的风物景观。武汉与水相生相伴，素有"江城"之称，又有"百湖之市"美誉。武汉因湖而美、因水而兴。武汉城市湖泊的环境、生态问题与流域社会、经济发展、城市扩张等密切相关。武汉作为全国水生态保护与修复、节水型社会"双试点"城市，政府十分重视对湖泊的管理，并随

东湖湿地　East Lake wetland

武汉墨水湖　Moshui Lake in Wuhan

着武汉河湖长制的有序推进，实施了一系列的湖泊污染防治和治理工程，湖泊水生态环境得到了显著改善，形成了多种武汉城市湖泊生态修复技术体系与模式实践。

1. 城市湖泊生态修复的"内沙湖模式"

武汉"内沙湖模式"是针对湖北大规模水产养殖造成"沉水植被退化消失，底泥深厚黑臭、水体严重污染"的湖泊生态现状，提出"以全面截污为前提，生态调查为基础，沉水植被构建为重点，鱼类群落调控为手段，生态系统结构调整为核心，兼顾景观"的具有湖北特色的城市退化湖泊生态系统修复新模式。

"内沙湖模式"在全面有效截污的基础上，通过原位生态改底，显著改善泥水界面氧化还原电位，有效消减内源污染；通过生物网膜技术和原位锁磷技术，快速实现水体透明度和营养盐浓度的下降，为沉水植被恢复提供良好基础条件。在此基础上，通过分步式沉水植物群落恢复技术，实现沉水植被的快速构建和水体水质的快速提升。最后，通过基于物质流和能量流的食物网构建，有效提升生态系统的物种多样性和生态系统稳定性，重建健康水体生态系统，恢复湖泊自净能力，实现湖泊生态系统的持久健康、稳定。

"内沙湖模式"注重湖泊底质改善，但不清淤、不挖泥，相比传统清淤治理，

东湖楚城　Chu city of East Lake

沙湖　Sha Lake

东湖　East Lake

内沙湖　Neisha Lake

投资少，见效快。该技术应用于武汉内沙湖水质改善治理，内沙湖已经连续9年稳定达到地表水Ⅳ类及以上标准。目前在全国多地已有多个湖泊使用这种模式进行修复。

"内沙湖模式"还在武汉市的都司湖、菱角湖、四美塘、小南湖、西北湖、后襄河、东湖、南湖、南太子湖、张毕湖、竹叶海、金湖、上金湖、墨水湖、杜公湖以及黄冈的遗爱湖和黄石大冶的尹家湖等湖泊应用，均取得显著成效。

2. 城市湖泊生态修复的"东湖模式"

武汉东湖是武汉最大的城中湖湿地，吹程长，风浪影响大，周边环境条件复杂，并承担重要的调蓄功能。这类湖泊直接开展生态修复，难度较大。

该模式是利用生态围隔，通过在东湖部分湖湾构建生态修复示范区，先行实现局部区域的生态修复，然后在听涛景区等重要岸线构建生态修复先行区，示范区与先行区连片后，可实现部分子湖的生态修复。武汉东湖通过"点—线—面"结合的方式，因地制宜，稳步推进各个子湖的生态修复，进而通过水生植被的自然扩增，最终实现整个东湖的生态修复。

目前，这种城市湖泊生态修复的"东湖模式"已在武汉南湖、汤逊湖、杜公湖及金银湖等湖泊推广应用，均取得了理想的效果。

3. 城市湖泊生态修复的"沙湖模式"

城市湖泊生态修复的"沙湖模式"是在"东湖模式"基础上研发的。武汉沙湖为过流型调蓄湖泊，汛期东湖湖水向西通过水果湖、楚河，进入沙湖，最终通过新生路泵站等外排入江。受老城区雨污分流改造工作难度大、持续时间久等影响，汛期承受大量合流雨污入湖污染负荷的冲击，沙湖具有水位波动大、水质和水生态系统受外部冲击大等特点。

"沙湖模式"采用生态围隔构建导流通道，汛期引导上游来水进入水果湖至楚河和沙湖大桥的围隔调蓄池，初步沉淀经新生路泵站外排入江，可保持围隔外生态修复区稳定。汛后，通过底质原位改善等措施，迅速提升导流通道内水体水质，促进通道内水生态系统的自我恢复与重建。

Chapter 4

Typical Cases of Wetland Conservation and Management in Hubei

In recent years, the province has ben conscientiously implementing Xi Jinping's thought of ecological civilization, firmly establishing and practicing the concept of "lucid water and lush mountains are invaluable assets", co-ordinating the system governance of mountains, water, forests, farmlands, lakes and grasslands, implementing the decisions and deployments of its Party Committee and government on wetland conservation, effectively performing responsibilities for wetland conservation and management, continuously promoting the conservation and restoration of wetlands in Hubei, and has yielded significant initial results.

On the basis of the comprehensive display of the practical achievements of construction of ecological civilization and in-depth introduction of the conservation achievements of important wetlands in Hubei, we summarize the exploration, innovation and demonstration experience of Hubei Province in wetland conservation and restoration, system construction and innovative management, rational use of wetlands and community building, science popularization and education and scientific research and monitoring. We have selected representative cases and innovative models of wetland conservation, restoration and comprehensive management in Hubei, and discussed the innovative and pioneering practices of wetland conservation and management in Hubei, while demonstrating the ecological beauty of wetlands and the harmony between man and nature in Hubei based on the ecological importance of wetlands, important wetland functions, conservation and scientific management, and science popularization and research.

4.1 Urban wetland conservation and integrated management in Wuhan, an international wetland city

4.1.1 Wuhan, the "city of a hundred lakes"

Wuhan, known as the "city of a hundred lakes" and the "city of wetlands". The Yangtze River, the largest river in China, and its largest tributary, the Hanjiang River, meet in the center of the city and are blessed with freshwater resources. The rivers in the city are crisscrossed with interwoven rivers and ditches and scattered lakes and reservoirs; the Yangtze River and Hanjiang River intersect in the center of the city and accept tributaries from the north and south into the confluence with 165 rivers of over 5 km and the existing water surface of

2,117.6 km². There are 166 lakes of all sizes, many of them on both sides of the Yangtze River, forming a network of lakes and marshes such as the well-known East Lake, South Lake, Tangxun Lake, Jinyin Lake and Shahu Lake, composing a very characteristic waterfront and lakeside water ecological environment.When the water level is normal, the lake surface area ranks first among cities in China, and makes Wuhan one of the megacities with the richest inland wetland resources in China, as well as a typical representative area of lake wetlands in the same latitude and middle and lower reaches of the Yangtze River in the world.

The wetlands in Wuhan cover an area of 162,000 hectares, accounting for 18.9% of the city's land area. 113,000 hectares of them are natural wetlands, accounting for 69% of the wetland area, and 49,700 hectares are artificial wetlands, accounting for 31%. The wetland resources rank among the world's top three in inland cities. The wetland conservation rate of the city is 53.19%. The wetlands in Wuhan are rich in biodiversity, inhabited by 413 species of wild animals including 38 species of national key protected wild animals: 10 species of national Class I protected animals such as *Ciconia boyciana* and *Ciconia nigra*, and 28 species of national Class II protected animals such as *Grus grus* and *Platalea leucorodia*. There are 408 species of vascular plants distributed on the wetlands, inlcuding 5 species of national Class II protected wild plants such as *Glycine soja*, *Nelumbo nucifera*, *Trapa incisa*, *Ceratopteris pteridoides* and *Ceratopteris thalictroides*.

As the kidney of city, wetlands in Wuhan play an important role in maintaining ecological stability, especially in reducing flood threats, weakening urban heat island and eliminating environmental pollution. To achieve sustainable urban development and play a good role as a central city in the central development strategy, Wuhan must reasonably use and conserve wetlands as an advantageous resource for urban construction.

4.1.2 Construction of wetland reserve systems in Wuhan

Since the establishment of Chenhu Lake Wetland Nature Reserve in 1994, wetland conservation in Wuhan has been going through nearly 30 years so far. Wuhan has been improving wetland conservation laws and regulations and system construction, reforming and innovating wetland conservation institutions and mechanisms, implementing major projects such as the renovation of the river bank, one lake one scenery, and the ecological water network of Dadong Lake, promoting the application of advanced technology of wetland conservation, mobilizing public participation in wetland conservation, and restoring the ecological environment of wetlands. At present, Wuhan has formed a regulation system of wetland conservation featuring lake conservation, and a wetland conservation system with nature reserves and wetland parks as the core, and the number of conserved areas is second to none among cities in China. There is one wetland of international importance-Chenhu Lake in Caidian, two provincial nature reserves - Chenhu Lake in Caidian and Zhangdu Lake in Xiinzhou, five national wetland parks – the East Lake, Jinyin Lake, Anshan, Houguan Lake, Dugong Lake and Canglong Island, four provincial wetland parks - Suozi Changhe and Tong Lake in Caidian, Zhuyanghai in Jiangxia, and Mulanhuaxiang in Huangpi, three municipal wetland nature reserves - Shangshehu Lake in Jiangxia, Caohu Lake in Huangpi and Wuhu Lake, and one national urban wetland park - Jinyin Lake.

Combining the migratory corridors of migratory birds in Wuhan area and the areas where the national key protected wildlife are concentrated to take water and feed, Wuhan has listed three places as key protected habitats for wildlife in the city including Qijiawan in Huangpi (*Aythya baeri*), Tianxingzhou in Hongshan (*Ciconia nigra*), and the Baiquan section of the Fu River in East and West Lake (Swan Lake).

4.1.3 Wetland conservation and management systems and innovative mechanisms in Wuhan

Wuhan was the first to incorporate wetland conservation and restoration into the legal system. In 1999, the Wuhan Municipal Government issued the *Measures on Conservation of Natural Mountains and Lakes in Wuhan*, placing equal emphasis on lake and wetland conservation and mountain conservation; and in 2007, the *Overall Plan of Wetland Conservation in Wuhan* was issued, incorporating wetland conservation into the overall plan of the city. In March 2010, Wuhan was the first sub-provincial city in the country to complete wetland legislation, promulgating the *Regulations on Wetland Nature Reserves in Wuhan*, designating a number of wetland parks and nature reserves and forming a Wuhan Sample for wetland conservation. In October 2013, Wuhan was the first city in China to launch the *Measures on Ecological Compensation of Wetland Nature Reserves in Wuhan*, a mechanism of wetland ecological compensation, investing more than 800 million yuan a year for ecological compensation to establish a mechanism of ecological compensation for wetland conservation. Wuhan launched the construction of the Intelligent Wetlands project, and was the first in China to compile the *Guide for Small and Micro Wetland Conservation and Restoration in Wuhan*. It has built the first batch of demonstration sites of small and micro wetland ecological restoration such as Baiquan, Dugong Lake and Tang Lake, increased the conservation effort of wetland nature reserves and habitats for wild birds such as Tianxingzhou, Qijiawan, and the Baiquan section of the Fu River and carried out surveys such Chenhu Lake wetlands of international importance and the biodiversity of Tianxingzhou, key protected terrestrial wildlife resources in the main stream of the Yangtze River, and wintering waterfowl.

By constructing multiple wetland conservation systems, Wuhan has made its riverbanks on both sides of the Yangtze River the largest waterside landscape in China, and the East Lake has been named the "most beautiful lake in the Yangtze River Economic Belt". It sets up a river and lake chief scheme to coordinate the upstream and downstream, left and right banks with a focus on the basin; 166 lakes in the city form the largest urban ecological wetland group of lakes in China; 165 rivers create enough waterfront space and green buffer zones.

After years of work, Wuhan has formed a wetland conservation synergy that is "government-led and department-coordinated with social participation", with more than 30 non-governmental organizations and 200,000 volunteers to help wetland ecological conservation. In addition, thanks to its wetland resources, Wuhan has established at least 12 educational venues and 69 bases for science popularization. It has built an exhibition hall for the achievements from 30th anniversary of China's implementation of the Ramsar Convention, carried out thematic publicity about World Wetlands Day, World Wildlife Day, and Bird Loving Week, compiled and published the *Seeking Dreams and Guarding Wetlands in Wuhan*, and made the documentary film *The City on the Wetlands* to show the achievements of wetland conservation in Wuhan and tell the story of its wetlands.

4.1.4 Implemented projects of important wetland conservation and restoration

In recent years, through wetland shoreline restoration, connection of river and lake water systems and restoration of wildlife habitats, Wuhan has continued to carry out a series of wetland conservation and ecological restoration work such as environmental improvement of the four banks of the two rivers, construction of the ecological water network of the Dadong Lake, connection of six lakes, four-water treatment, pilot sponge city and has gained remarkable results.

Project of Wetland Conservation and Restoration of Chenhu Lake: Through dredging the water system, conserving native vegetation, constructing diversified bird habitats and establishing an intelligent wetland supervision model, the ecosystem of the wetlands has been restored, the conservation and management capacity has been improved, and the ecological environment of the wetlands in Chenhu Lake has been continuously improved.

Project of Green Island Station Renovation and Upgrading of the East Lake.

Project of Wetland Restoration of Zhangduhu Lake: Wuhan has planted Chishan forest on the water forest, planted aquatic plants, constructed bird conservation fence and bird watching walkway to comprehensively improve the environment along the water shoreline.

Project of wetland conservation and restoration of Dugong Lake: It has carried out projects such as returning land for farming to wetlands, returning land for residence to wetlands, returning land for construction to wetlands, marsh wetland restoration, ecological water replenishment, lake improvement, vegetation restoration and habitat improvement.

4.2 Management exploration on protected areas of rare species in wetlands of Tian-e-zhou Oxbow of the Yangtze River

4.2.1 Special characteristics and typicality of the wetlands in oxbows in the middle reaches of the Yangtze River

The Jing River is a section of the middle reaches of the Yangtze River in the alluvial plain. Influenced by topography, geology and hydrology, at the Jingzhou section of the Yangtze River featuring deep and fast flow and frequent bank collapses, the natural changes of the river channel as well as man-made influences led the original curved river channel to cut straight, and the upper and lower mouth gradually silted and closed, forming the oxbow of the Yangtze River (ox yoke lake). A large wetland has formed by the shoal of the oxbow of the Yangtze River. Tian-e-zhou oxbow is the natural bend of the Yangtze River formed in 1972. Originally, it was the oxbow of the Tongjiang River. In 1998, a dike was built in Shatanzi to block the Yangtze River, making the oxbow connect to the Yangtze River only through the downstream Tian-e-zhou gate.

Within the area of the wetlands in Tian-e-zhou oxbow, there are Hubei Shishou Milu National Nature Reserve and Hubei Changjiang Tian-e-zhou Baiji National Nature Reserve, which protect the Yangtze *Neophocaena asiaeorientalis* and Elaphurus davidianus respectively, as well as their habitats. In addition, the wetlands are home to *Ciconia boyciana* - a IUCN endangered species, *Anser cygnoides*, *Anser erythropus*, *Calidris tenuirostris* and *Emberiza aureola - vulnerable* species, *Aythya baeri* – a critically endangered species, *Falco amurensis*, *Coturnix japonica*, *Limnodromus scolopaceus*, *Limosa limosa*, *Numenius arquata*, *Corvus torquatus* and *Emberiza yessoensis* - near-threatened species (NT). The typical wetland ecosystem in the Tian-e-zhou oxbow of the Yangtze River has a complete ecological structure, unique functions and rich wildlife resources, which is important for the conservation of the typical unique wetland ecosystem and species diversity. The Tian-e-zhou oxbow of the Yangtze River combines natural geography, rare animals and conservation of natural wetland ecology, showing important value in scientific research, culture, ecology and tourism.

4.2.2 Construction and management of national nature reserves

4.2.2.1 Hubei Shishou Milu National Nature Reserve

Hubei province with a total area of 1,567 hectares. It is set up to restore the natural population of *Elaphurus davidianus* in its native habitat and to protect the ecological environment of the wetlands on which it lives. Through years' efforts of the conservation staff, the population of *Elaphurus davidianus* in Shishou Milu National Nature Reserve has increased rapidly from 64 heads introduced from Nanhaizi Milu Park in Beijing in October 1993 and November 1994 in two batches to more than 2,500 today; four sub-populations have been formed in the core area, Jiangnan Sanheyuan, Yangbotan in Xiaohe Town and Dongting Lake in Hunan province, and all of them have been left to grow in the natural environment and released to the wild.

Expanding space for activities of *Elaphurus davidianus*. In order to solve the conflict between human and *Elaphurus davidianus* for land, with the support of the Party Committee and government, the land ownership of 15,000 mu in the core area of protected area and experimental area as well as land use rights for 8,000 mu in the buffer area have been implemented, and the space for activities of *Elaphurus davidianus* has been gradually expanded. By strengthening the control in the protected area, the peaceful environment within the protected area is ensured.

Creating favorable living conditions. First, food security. Under the premise of strengthening the habitat conservation of Tian-e-zhou, the emergency feed base for *Elaphurus davidianus* has been built. Second, water souce security. Water exchange is carried out through the construction of drainage pumping stations and pipe networks to ensure clean water sources. Third, secluded sites. The reserve has a large area of reed and peking willow, which can provide a good hiding place for *Elaphurus davidianus* to give birth and avoid the wind.

Carrying out biodiversity monitoring. The reserve focuses on *Elaphurus davidianus* monitoring with full utilization of new technologies and equipment such as 3S, drones and infrared cameras. Through the monitoring of their habits, activity patterns and environmental needs, we can also provide scientific basis for the conservation, restoration and management of the habitats of *Elaphurus davidianus*.

Managing habitats well. The habitat management is carried out based on the living habits of the *Elaphurus davidianus*, and the guaranteed artificial grazing area is built to meet the needs of them for safety during flood period.

Strengthening prevention and control of epidemics and its source and scientific research cooperation. The reserve strengthens its cooperation with scientific research units and colleges and universities to identify the weak links of prevention and control of epidemics and its source in the reserve, and establishes and improves a system for early warning, prevention and control for epidemics of *Elaphurus davidianus* as well as a monitoring platform for scientific research.

4.2.2.2 Hubei Changjiang Tian-e-zhou Baiji National Nature Reserve

In 1992, the Hubei Changjiang Tian-e-zhou Baiji National Nature Reserve was established, and the researchers tried to use the Tian-e-zhou oxbow of the Yangtze River to relocate and protect *Lipotes vexillifer* (now extinct) and Yangtze *Finless Porpoise*- the endangered species of the Yangtze River. Since the release of five *Finless Porpoise* in the nature reserve, a population of 101 has been developed after a census in 2021, and 6-8 young *Finless Porpoise* are born every year, laying a firm foundation for the recovery of the natural population of Yangtze *Finless Porpoise*.

The Tian-e-zhou oxbow of the Yangtze River has become a protected area with the largest population of Yangtze *Neophocaena asiaeorientalis*, and Shishou city has been called " the hometown of Chinese *Neophocaena asiaeorientalis*". The ex-situ conservation of the *Neophocaena asiaeorientalis* has been hailed by domestic and foreign experts as the only successful example of ex-situ conservation of cetaceans in the world. Under the circumstance of a sharp declining of the overall population of Yangtze *Neophocaena asiaeorientalis*, the ex-situ conservation of the *Neophocaena asiaeorientalis* in the Tian-e-zhou oxbow has brought hope for the survival and conservation of the *Neophocaena asiaeorientalis* Porpoise and provided valuable experience for the conservation of other rare and endangered aquatic animals in the Yangtze River. In February 2021, with the approval of the State Council, National Forestry and Grassland Administration and the Ministry of Agriculture and Rural Affairs announced the newly revised *List of National Key Protected Wildlife*, in which the Yangtze *Neophocaena asiaeorientalis* is upgraded to the national Class I protected wildlife. The *Neophocaena asiaeorientalis*, the flagship species of the Yangtze River and the "smiling angel" from nature, will swim in the Yangtze River under the careful care of people.

4.2.3 Conservation of wetlands in oxbows and ex-situ conservation of endangered species

In the wetlands of Tian-e-zhou oxbow, the entire wetland ecosystem is conserved mainly by establishing protected areas and introducing endangered and rare species such as *Elaphurus davidianus* and *Neophocaena asiaeorientalis*, and ex-situ conservation is often a very important and necessary conservation meas for species with rapidly declining populations.

4.2.3.1 Conservation of Yangtze *Neophocaena asiaeorientalis*

After 30 years of development, the conservation of the Yangtze *Neophocaena asiaeorientalis* in the Tian-e-zhou reserve has been effective, and the reserve has been recognized by experts from home and abroad as the only successful example of ex-situ conservation for a cetacean. In 2021, the population of *Neophocaena asiaeorientalis* in the Tian-e-zhou oxbow has exceeded 100 with an annual growth of about 10%, reaching the environmental capacity of the population. Now the Tian-e-zhou oxbow is the only area with a stable growth in the population of *Neophocaena asiaeorientalis* within the Yangtze River basin.

The Yangtze *Neophocaena asiaeorientalis* is being protected in ex-situ conservation. In the 89-km main stream of the Yangtze River in Shishou in the reserve, through cracking down illegal activities, improving environment comprehensively and restoring ecology, more than 20 individuals are now inhabiting in the Shishou section of the Yangtze River year-round, and they often appear in groups.

The Yangtze *Neophocaena asiaeorientalis* is cultured in net pens. The net-pen culture of porpoise and net-pen breeding technology in the reserve were awarded national invention patents. Beibei, the first porpoise cultured in net pens was released into the Tian-e-zhou oxbow in July 2020 when it reached the age of four through rewilding training.

The gene exchange of *Neophocaena asiaeorientalis* population is promoted in a scientific manner. In order to improve the genetic structure of porpoise population in the Tian-e-zhou oxbow and prevent the genetic decline of the population, the Tian-e-zhou reserve has exported 24 *Neophocaena asiaeorientalis* to reserves such as Xinluo section of the Yangtze River, Hewangmiao in Jianli, Tongling in Anhui and other suitable habitats since 2015, while 8 *Neophocaena asiaeorientalis* have been introduced from Poyang Lake of Jiangxi. According to the *Action Plan of Rescue of Yangtze Neophocaena asiaeorientalis*, the reserve will continue to export and introduce Yangtze *Neophocaena asiaeorientalis* to improve their population genetics.

4.2.3.2 *Elaphurus davidianus* and its habitat conservation

Through effective conservation, the biodiversity of the reserve has been improved and the population size of the *Elaphurus davidianus* has been expanded. According to the latest comprehensive scientific examination, the number of higher plants in the reserve has increased from 238 in 1997 to 321, the number of terrestrial vertebrates, from 90 in 1997 to 320, and the number of national key protected species has reached 56. Especially the rare and endangered birds such as *Ciconia nigra*, *Ciconia boyciana*, *Ciconia ciconia* and *Antigone vipio* have been found in recent years.

The number of *Elaphurus davidianus*, the key protected species, has increased from 64 to more than 2,500, and four sub-populations have been formed in the core area, Jiangnan Sanheyuan, Yangbotan in Xiaohe Town and Dongting Lake in Hunan province, and all of them have been left to grow in the natural environment and released to the wild. The reserve has been jointly commended by seven national ministries and commissions as National Advanced Collective of Nature Reserves twice, and has been ranked as excellent in the management assessment of national nature reserves twice in a row.

4.3 Innovation exploration on wetland conservation and management in Shennongjia National Park

4.3.1 Ecological positioning and regional wetland characteristics of Shennongjia National Park

Shennongjia is the only well-preserved primitive area with forest distribution in the mid-latitude region on the earth. It enjoys unique natural environment, diverse, ancient, relic and rare species, rich water resources, as well as unique tourism resources and rare plant and animal resources. Known as a "natural museum", "bank of species gene", and "refuge for relict plant and animal", it is an ideal place for the survival and reproduction of wild animals and plants. Thanks to its intact forest ecosystem and abundant precipitation, it is an important water-conserving area in the Yangtze River basin, and plays an important role in improving the ecological environment and maintaining ecological balance.

The wetlands of Dajiu Lake in Shennongjia are located at the northwest end of Hubei, situating on the watershed of the Yangtze River and Hanjiang River, bordering Chongqing and Shaanxi to the west, Shiyan city in Hubei to the north, and Badong county in Enshu to the east, which is known as "one foot in three provinces and six counties". With the average altitude of 1,730 meters, the wetlands of Dajiu Lake are surrounded by the Shennongjia mountains, forming a unique alpine basin. They form a unique natural landscape with the basin and mountains, and the natural forest vegetation and vegetation in wetland marshes are intertwined in a charming painting. The Dajiu Lake in Shennongjia is a perfect combination of wetland ecosystem and forest ecosystem and the unique wetland structure makes the wetlands here more diverse than other wetlands. Rare animals such as *Rhinopithecus roxellana*, *Aquila chrysaetos* thrive here; relict plants represented by *Davidia involucrata*, *Liriodendron chinense*. and *Cercidiphyllum japonicum* are sheltered in this special natural environment. Such wetlands are even extremely rare in China, and were included in the list of wetlands of international importance by the Ramsar Convention in 2013.

The Third Plenary Session of the 18th CPC Central Committee made a decision on strengthening the construction of ecological civilization and clearly proposed the establishment of a national park system. In November 2016, Shennongjia entered the implementation phase of the pilot national park system. The total area of the pilot area of Shennongjia national park system is 1,170 square kilometers, accounting for 35.97% of the total area of Shennongjia forest area, and the forest coverage rate of the pilot area is up to more than 96%. The Shennongjia national park is located at the eastern edge of the second terrain of China's topography, and belongs to the middle and high mountains of the eastern extension of the Daba Mountain Range, one of the priority areas for biodiversity conservation in China. It is the watershed between the two mother rivers, the Yangtze River and Hanjiang River. The Duhe River system of the Shennongjia mountains, with a basin area of 791 square kilometers, reaches the Danjiangkou reservoir after converging into the Hanjiang River, and accounts for 20% of the water volume of the Danjiangkou reservoir. Therefore, with a very important ecological status, Shennongjia is the ecological cornerstone of the green development of the Yangtze River Economic Belt, the important water conservation area of the medium route project of the South-to-North Water Diversion, and the largest natural green barrier in the Three Gorges reservoir area. It is related to the ecological security of the country and sustainable economic development.

4.3.2 Ecological construction and wetland conservation in Shennongjia

Since the start of the pilot work of Shennongjia national park system, it has been practicing Xi Jinping's thought of ecological civilization and the concept of "lucid waters and lush mountains are invaluable assets", adhering to the concept of national park of ecological conservation first, national representation and public welfare for all. Aiming at conserving the originality and integrity of natural ecosystem, it coordinates the integrated conservation of mountains, rivers, forests, farmlands, lakes and grasslands, and actively explores the new path of wetland conservation and management of the Dajiu Lakes, improving the ecological environment, water quality at the source and wetland ecological functions significantly.

Rationalizing management system and cohesion of conservation. The Shennongjia natural reserve implements conservation and management in the manner of zoning, grading, and classification, converting from conservation of a single forest ecosystem to an integrated conservation of mountains, rivers, forests, farmlands, lakes and grasslands.

Adhering to the norms and strengthening the legal protection. The *Regulations on Shennongjia National Park Conservation* were promulgated to implement "one park, one law" management; the Positive and Negative List of Industry Access Into Shennongjia National Park was issued, forming a more complete system of regulations and institutional norms.

Carring out pollution prevention and control and fully implementing river and lake chief scheme. The Shennongjia natural reserve adheres to the "one river, one policy" and "one lake, one policy" to timely solve the key and difficult problems that restrict the water environment, leading to a total of nine small hydropower stations being shut down.

Strengthening ecological restoration and conserving important wetlands. In order to ensure that "a reservoir of clean water is sent north forever", the overall ecological relocation of 460 former residents in the wetlands of the Dajiu Lake has been carried out since 2013; no-grazing areas have been designated in the core area of the wetlands, to eliminate the possibility of large-scale non-point source pollution. The Shennongjia natural reserve also explores new ideas for the conservation and restoration of peat moss mire wetlands.

Building a three-dimensional network and realizing intelligent management. The information management center of Shennongjia national park and the early warning system for tourist flow in scenic spots have been built; monitoring means such as satellite remote sensing, drones and infrared cameras have been used to form a three-dimensional conservation network and achieve a full coverage of biological monitoring.

4.3.3 Innovative mechanisms of community co-construction in ecological conservation

The Shennongjia national park explores a sustainable model of conservation and management and community co-construction to consolidate the conservation achievements. Since November 2016, a total of 146 million yuan has been invested in community co-construction and co-management, industrial development, infrastructure construction, ecological compensation and targeted poverty alleviation in the five townships under the jurisdiction of the national park.

Establishing linkage mechanism of community co-construction and co-management. The Shennongjia national park signed cooperation agreements of community co-construction and co-management with 25 administrative villages in 5 townships under its jurisdiction, setting up working institutions and a sound contact mechanism to realize the development goals of community co-construction and co-management of ecological conservation, industrial development, capacity enhancement, social management and sharing of reform results in the national park communities.

Actively promoting ecological relocation and carrying out ecological compensation mechanism. The Shennongjia national park firmly establishes the concept of co-construction and sharing, and focuses on establishing a mechanism for benefit sharing and coordinated development. Since the implementation of ecological relocation, a total of 976 million yuan has been invested in compensation for ecological relocation and the construction of the resettlement area of Pingqian ancient town. It vigorously implements the pilot subsidy project of "electricity instead of fuel" in accordance with the standard of 3,000 yuan per household per year for ecological migrants, changing the previous way of cutting down forests and burning wood for heating, so as to achieve the purpose of maximizing the conservation of the forest resources in Shennongjia while ensuring the residents' winter heating, and minimizing the threat to ecological resources from residents' use of firewood for production and domestic life.

4.4 Practices of lake conservation and integrated governance in wetlands of international importance in Honghu Lake

4.4.1 Ecological location and functions of wetlands in Honghu Lake

The classic song of "Honghu Waters, Wave upon Wave, the Shore of Honghu Lake is Home ......" has become the bright business card of Honghu Lake. Honghu wetlands are located in the southeast of Hubei Province, the north bank of the middle reaches of the Yangtze River, the downstream of the four-lake basin in the Jianghan Plain, straddling the cities of Honghu and Jianli. In 2012, "one lake, one survey" determined the surface area of Honghu was 308 km^2, making it the largest lake in Hubei Province and one of the largest lakes in the middle and lower reaches of the Yangtze River. Honghu Lake is an important lake mainly for storage, but also for irrigation, fishing, shipping and biodiversity conservation, and its wetlands were included in the list of Wetlands of International Importance in 2008, and promoted to a national nature reserve in December 2014.Honghu wetlands are included in the list of important wetlands in China, a priority conservation project of action plan of wetland conservation in China, and one of the 238 most important ecological zones in the world as determined by WWF.

Honghu Wetland Nature Reserve enjoys beautiful scenery, rich biodiversity and well-maintained wetland ecosystem. According to the survey, there are 472 species and 21 varieties of vascular plants and 1 form in the reserve, and 8 of them are national key protected plants. There are 708 species of animals including 231 species of vertebrates, 62 species of fish, 138 species of birds, 6 species of amphibians, 13 species of beasts, 379 species of zooplankton and 98 species of benthic animals. And 24 species of vertebrates are nationally key protected; the national first grade key protected animals are *Ciconia boyciana*, *Ciconia nigra*, *Mergus squamatus*, *Otis tarda*, *Aquila heliaca* and *Haliaeetus albicilla*.

Honghu wetlands belong to the nature reserve of inland wetlands and freshwater ecosystem type, whose main conservation objects are permanent wetland ecosystem of freshwater lakes, freshwater resources and rare waterfowl, as well as protecting the freshwater resources and biodiversity of Honghu Lake, maintaining the structure of lake ecosystem and functions of ecological environment, so that Honghu wetlands can permanently, continuously and effectively serve the coordinated development of regional economy and society, and promote the regional construction of ecological civilization.

4.4.2 Conservation, restoration and integrated management of wetlands in Honghu Lake

Honghu Lake, located in the middle reaches of the Yangtze River in the Jianghan lake group, is an important ecological barrier in Jingchu. In recent years, Hubei Provincial Party Committee and Provincial Government and Jingzhou Municipal Party committee and Municipal Government have made various decisions and arrangements under the guidance of "promoting well-coordinated environmental conservation of the Yangtze River", and actively done a good job of seine-net removal, fishermen's landing, and ecological restoration with the focus on the conservation and restoration of Honghu wetlands, to achieve a "win-win" of ecological conservation and economic and social development. In April 2018, when General Secretary Xi Jinping inspected Hubei, he personally visited Jingzhou to feel the pulse of the development of the Yangtze River Economic Belt and navigate the development of Hubei and Jingzhou, which is of milestone significance. The wetland conservation of Honghu Lake focuses on grasping the three relationships and win the "three" battles to continuously improve the ecological environment.

Handling carefully the relationship between overall promotion and key breakthroughs to win the battle of removing seine-net. The sentence of "Lucid waters and lush mountains are invaluable assets." is realized through the real action of Honghu Lake. Honghu Lake is the largest gathering of professional fishermen in Hubei. The province vigorously developed its aquaculture in the 1980s with the seine-net area up to 377,000 acres in its prime in Honghu Lake, accounting for 71% of the lake area, which formed an overuse causing serious damage to the ecological environment. The ecological environment of Honghu Lake is highly valued by the successive Provincial Party Committees and provincial governments, and they launched a 12-year battle to remove the seine-net. In the process, with the full involvement of the government, the city, county, township and village levels linked together to implement compensation measures, and fishermen cooperated deeply, which created the "Honghu speed of removing seine-nets", and also the first in the country in terms of the one-time removal of the seine-net area of a single lake.

Handling carefully the relationship between ecological environment conservation and economic development to win the battle of fishermen's landing. The seine-net removal is only the first step in the ecological environment conservation of Honghu Lake. The fundamental solution to eradicating illegal seine-nets and fishing in Honghu Lake lies in the landing of fishermen and their conversion to other industries. There are 3,512 households of fishermen and 12,295 people affected by seine-net removal in Honghu Lake, including 526 households and 1,902 people who have been relocated to alleviate poverty (2016-2017), allowing a new chapter of the fate of fishermen to be opened. The Honghu Wetland Nature Reserve adheres to the idea of "relocation for poverty alleviation as the basis, supporting employment and entrepreneurship as the direction, sound social security as the bottom line", and eliminates the fishermen's worries. By upgrading the aquatic industry around the conservation area, the reserve ensures the status of Honghu city as the first county-level city of freshwater fishery in China. Through the improvement of the environment, the construction of the greenway around the lake, the development of ecological tourism, the local economy has been injected new vitality to achieve transformation and upgrading.

Handling carefully the relationship between overall planning and long-term efforts to win the lasting war of ecological restoration. Honghu Lake, carrying more than 10,000 square kilometers of upstream water from the four-lake basin of the Yangtze River, is known as the "kidney of the Jianghan Plain". As an important part of the well-coordinated environmental conservation of the Yangtze River, the conservation of Honghu Lake is a great responsibility. Targeting a "smooth lake, clear water, green shore, beautiful scenery", Jingzhou Municipal Government puts forward the idea of "basin governance, comprehensive governance, project governance, responsibility governance", adheres to the planning priority to ensure the effectiveness of governance of Honghu Lake. It advances comprehensive management and establishes the four-level lake chief scheme consisting of 17 lake chiefs and the joint meeting system consisting of the government and municipal departments. It upholds the comprehensive governance of the entire basin. Specifically, the entire basin 14 enterprises of "ten small" industries with serious pollution have been closed; 54 drain outlets have been blocked; 14 enterprises of printing and dyeing and paper manufacturing have been transformed in energy-saving; 70 large-scale poultry farms and 310 under-scale farming professionals have all been shut down and relocated. Comprehensive measures have made the ecological governance of Honghu Lake effective.

4.5 Danjiangkou reservoir, and important water source and conservation of its basin ecology

4.5.1 Wetlands and water resources in Danjiangkou reservoir area

The Hanjiang River is the largest tributary of the Yangtze River, and its reservoir basin has a well-developed water system with many rivers, all belonging to the Hanjiang River system, the first major tributary of the Yangtze River. The rivers all flow from the north to the south to Danjiangkou reservoir. The main rivers are the Danjiang River, Laoguan River, Qihe River and Taohe River, and most of them have deep channels, large slopes and rapid currents.

The current water quality of Danjiangkou reservoir meets standard II of the water environmental quality standards of surface water, and the requirements of various functional uses of water. Danjiangkou reservoir area and its upstream area are the water source area of the medium route project of the South-to-North Water Diversion, which is a national strategic zone for water resources.

The wetlands in Danjiangkou reservoir are one of the representative wetland types in China, a nature reserve of inland wetlands and water ecosystem type with river-type reservoir ecosystem as the main conservation object. In 2000, they were included in the implementation scope of priority action of the *Action Plan of Wetland Conservation in China* and the list of important wetlands in the southwest region.

The South-to-North Water Diversion Project is a major strategic initiative to optimize the allocation of water resources in China. As a core water source area of the Project, a national strategic zone for water resources, a demonstration area for the conservation and development of inland river basins, a pilot area for linkage development in central and western China and a pioneer area for green development in the Yangtze River basin, the Danjiangkou reservoir area in the Hanjiang River Basin has the function of "integrating two circles, connecting one belt, linking the north and the south, and carrying the east and the west". It has an important strategic position and plays a prominent driving role in the pattern of quality development of the Yangtze River Economic Belt. As of May 2021, the Danjiangkou reservoir, known as the "Water Capital of China", has transferred 37.9 billion cubic meters of water to Beijing, Tianjin, Hebei and Henan, accounting for 87.9% of the total water transferred of the South-to-North Water Diversion Project, benefiting about 79 million people in 24 large and medium-sized cities and bringing into full play its social and economic benefits.

4.5.2 Water source conservation and ecological compensation in Danjiangkou reservoir area

Water source conservation and water security. Since 2018, the national pilot program of "returning land for farming to wetlands" has been implemented in Danjiangkou reservoir area with adherence to the guidelines of "comprehensive conservation, scientific restoration, rational utilization and sustainable development" and the basic principles of "integrated planning, concentrated contiguity, optimized layout, exemplary guidance, and gradual advancement". The area seizes the opportunity to carry out project tasks such as planting and replanting, ecological water replenishment, returning land for farming to grasslands and returning land for farming to wetlands. 14,000 mu of "returning land for farming to wetlands" will be fully completed by the end of 2020, including artificial afforestation, artificial grass planting, ecological water replenishment, and closing hillsides to facilitate afforestation. The implementation of these projects has improved the forest coverage of the reservoir area, further enhanced the water conservation and soil conservation capacity of the wetlands, optimized the dominant function of the wetland ecosystem, improved the ecological environment of the water source area, provided a strong guarantee for the ecological security of the South-North Water Diversion Project, and made an important contribution to the transfer of quality and clean water resources to the north for the South-North Water Diversion Project.

"Full green in Jingchu" and "green shield" action. The construction of water-conserving forests is an important way to improve the ability to conserve water sources and the safety of water quality in water source areas. In recent years, the land greening in Shiyan Hubei where the reservoir area is located, based on the ideas of greening of barren hills, greening of the gateway, and greening of the channel, with the Hanjiang River and the Duhe River as the focus, carries out a solid afforestation at "waterside, roadside, village side", completes the major projects of "full green in Jingchu" and "targeted anti-desert", implements the special campaign of afforestation on both sides of the Yangtze River and the Hanjiang River, continues to promote the projects of protection forests and returning land for farming to forests, restoration and treatment of bare mountains and forest form improvement to increase aesthetic value, laying a solid foundation for the ecological security of the reservoir area and water security. At the same time, the reservoir area carries out special action of "green shield" to strengthen the supervision of ecological environment.

Mechanism of ecological compensation and development of ecological industry. Putting the "a reservoir of clean water is sent north forever" at the overriding position, the reservoir area actively explores the implementation of the mechanism of ecological compensation, advances the procedures of "pre-approval, mid-event supervision and post-monitoring" for major engineering projects when they need to involve nature reserves, implements the mechanism of ecological compensation to ensure the ecological safety. It also gives full play to the social responsibility of enriching forestry and the people, promotes the industrial structure adjustment and efficiency improvement in the reservoir area, cultivates leading industries according to local conditions, and realizes the win-win situation of ecological construction and industrial development in the reservoir area. The reservoir area vigorously stimulates rapid expansion of industries with special characteristics such as woody oilseeds, seedlings and flowers, forest economy and forest tourism, as well as the rapid development of the Olea europaea industry around the reservoir with the total scale exceeding 50,000 mu, having an important impact inside and outside the province. The golden oil-producing belt of 32° north latitude has taken shape.

4.6 Integrated efforts for wetland conservation of the Hanjiang River in Xiangyang, collaboration mechanism of basin conservation as a pioneer

The Hanjiang River is an important underpinning for the ecological pattern in Hubei and is the mother river of Xiangyang. It meanders for 920 km within the province. With the continuous promotion of construction of ecological civilization, the Hanjiang River in Xiangyang has taken a green development path of holistic conservation, systematic restoration and comprehensive governance, presenting a clearly visible beauty of ecology, intelligence and harmony and adding bright colors to the great beauty of wetlands in Hubei.

The beauty of the Hanjiang River in Xiangyang lies in the clear water of a river. The water quality of the Hanjiang River is the best among the big rivers. According to the monitoring data, the overall water quality is stable above the second grade with some indicators close to or exceeding the first grade water quality, which is similar to the water quality in the central area of Danjiangkou reservoir. The open water surface and clear water bodies, together with more than 200 historical sites and numerous legends of the River around the area, are the greatest features of the Hanjiang River ecology, harboring the beauty of intelligence.

The beauty of the Hanjiang River in Xiangyang lies in the water birds chasing the waves. As an important migratory passage and station for migratory birds, the waterfowl resources of the river have been significantly restored as the ecological environment has improved. The number of waterfowl species has increased from 78 five years ago to 132 now, and the population size has increased from more than 90,000 to more than 160,000, with a few hundred or thousands of waterfowl flying and ducks swimming around, showing the beauty of harmony.

The beauty of the Hanjiang River in Xiangyang lies in its green shores. Clean water and green grass ripple along the 195-kilometer tree-lined shores of the river. The shores of the Hanjiang River are naturally extended with green clothing, and the shoal and islands at the center are of different shapes and forms, which are naturally formed and full of life. More than 500 species of native plants are used to build beautiful homes for wetland birds and other animals. The effect of the ten-year ban on fishing has been seen, and fish are chasing and swimming at will. This is the original beauty of the wetlands of the Hanjiang River.

4.6.1 An innovative collaborative mechanism for the conservation of the Hanjiang River, the key to wetland conservation of the Hanjiang River

To better conserve the wetlands, Hubei Province focuses on the big basin and has established Alliance of Wetland Conservation of the Hanjiang River (Xiangyang section in Hubei), the first wetland conservation alliance in Hubei Province, at the Xiangyang section of the Hanjiang River basin. It was initiated by the Hubei Wetland Conservation Foundation and established in consultation with six national wetland parks along the Hanjiang River (Xiangyang section in Hubei). The alliance follows the path of "pilot first, gradual improvement, and collaborative development", carries forward the spirit of "common mission, common vision, and joint responsibility", builds a new conservation pattern for the wetlands of the Hanjiang River, implements the newly promulgated wetland conservation law better, integrates cooperation, sharing and innovation among alliance units in the Xiangyang area of the Hanjiang River basin, promotes scientific conservation, information sharing, upgrading and joint prevention and management, ensures the ecological functions of the wetlands of the Hanjiang River, maintains biodiversity and more stable ecological security, and finally achieves a deeper implementation of the construction of ecological civilization of the Hanjiang River basin, and the Hanjiang River model effect of harmonious coexistence between human and nature.

Scientific conservation. The alliance should establish demonstration bases for quality conservation and restoration of wetland parks, reform and innovation, and promote typical examples to achieve the leading role of models and overall promotion.

Information sharing. All members of the alliance should strengthen the common construction of science popularization and education, scientific research and monitoring, and network platforms to realize the interconnection of network platforms, and sharing of information and results.

Upgrading. The alliance should promote environmental monitoring and resource survey of the Hanjiang River to provide scientific basis for wetland conservation and science popularization and education.

Joint prevention and management. The members of the alliance should jointly carry out actions to conserve migratory birds flight and water quality of the Hanjiang River.

4.6.2 Maintaining connectivity and mobility of the Hanjiang River water system, the source of the health of the wetlands of the Hanjiang River

The Hanjiang River basin will implement key projects such as the nine waters nourish city, the restoration of ancient Xiang Water, and the oxbow restoration of the Behe River, carry out the connectivity project of the Hanjiang River water system in the whole area, cope with the six-levels hydropower development in the Hanjiang River in the province, and implement the unified deployment of floods and water volume to ensure that the wetlands of the Hanjiang River are abundant in water, healthy and smooth with long-lasting water currents.

4.6.3 Taking environment improvement actions such as ten major campaigns, the basis of wetland security of the Hanjiang River

The Hanjiang River basin will build a work pattern to protect the Hanjiang River in a well-coordinated manner and carry out ten major campaigns such as the special rectification of chemical enterprises along the river, the rectification of agricultural non-point source pollution, the rectification of illegal docks, the rectification of illegal sand mining, and the reduction of sewage from enterprises along the river, to provide a basic guarantee for a beautiful ecological environment of the wetlands of the Hanjiang River.

4.6.4 Restoring and improving the original ecology of the Hanjiang River, the development strategy for wetlands of the Hanjiang River

The Hanjiang River basin will systematically carry out the returning of land for farming to wetlands, the returning of land for breeding to wetlands, the demolition of illegal buildings in wetlands, the people retreating and wetlands advancing to implement a ten-year ban on fishing to restore the food chain structure of the Hanjiang River, focus on implementing restoration projects of bird habitats, promoting restoration of native vegetation, and effectively improving the quality and stability of the wetland ecosystem of the Hanjiang River.

4.6.5 Creating a green space shared by the people, the way to benefit the people of the wetlands of the Hanjiang River

The six national wetland parks of the Hanjiang River in Xiangyang are constantly increasing their construction efforts to create a beautiful ecological environment, provide excellent ecological products and quality ecological services, and organically combine quality nature education, study bases and ecological recreation to make themselves ideal places for citizens to relax and embrace nature.

4.7 Wetlands of international importance of Chenhu Lake, intelligent wetlands with a lifted level of scientific management

4.7.1 Wetlands of international importance and the nature reserve in Chenhu Lake

A district-level nature reserve was established in the wetlands of Chenhu Lake in January 1994, and promoted to the municipal level the following year, to the first provincial wetland nature reserve in Wuhan in August 2006 and was identified as wetlands of international importance in October 2013. The wetlands of international importance of Chenhu Lake (Chenhu Wetland Nature Reserve at the provincial level in Hubei) is located in the southwest of Caidian district, Wuhan, and is an inland wetlands and water ecosystem type nature reserve with typical wetland ecosystems and rare waterfowl as its main conservation objects. It mainly consists of Chenhu Lake, Zhangjia First Lake, Wangjiashe Lake and part of Dujiatai flood storage area, which is the largest freshwater lake and flooded marsh wetlands of the Jianghan Plain.

The reserve is rich in wildlife resources, with 330 species of vascular plants including 4 species of national Class II protected plants. There are 73 species of benthic animals, 58 species of fish, 10 species of amphibians, 28 species of reptiles, and 27 species of mammals. There are 277 species of birds including 14 species which are Class I protected at the national level, 50 species, Class II protected at the national level, and 47 species, protected at the provincial level. More than 100,000 migratory birds rest, breed or winter here every year. The distribution number of 18 species of birds exceeds 1% of the total global population, which is regarded by experts as "the genetic depository of wetland waterfowl".

4.7.2 Biodiversity monitoring and construction of intelligent wetlands of Chenhu Lake

4.7.2.1 Biodiversity monitoring of Chenhu Lake

Chenhu Lake is an important bird habitat with rich biodiversity. The reserve actively cooperates with scientific research units and environmental conservation organizations such as Bird Watching Society in Wuhan to carry out special monitoring of wintering birds and regular bird monitoring, monitoring and assessment of wetland biodiversity, survey and monitoring of wetland ecological environment and geology, and prevention and control of epidemics and its source of wetland wildlife, so as to provide scientific decision-making basis for the conservation and management of the wetlands in Chenhu Lake.

4.7.2.2 Construction of intelligent wetlands of Chenhu Lake

The construction of biodiversity monitoring and intelligent wetland platform of Chenhu Lake is based on the basic principles of ecology and conservation biology, and the system solution created by the collection of the most advanced artificial intelligence, Internet of Things, big data, and cloud computing to explore real-time analysis of various ecological data and ecological early warning, which is the first intelligent wetland platform in Wuhan.

Biodiversity sensing network. The design and construction of a more complete sensing network covers the monitoring needs of different ecological elements of different scales of the sky, water, soil, atmosphere and organism.

Wetland ecological brain. It includes various modules such as realistic dynamic simulation of the wetlands in Chenhu Lake, video recognition of wildlife, vocal recognition of wildlife, intelligent interpretation software of remote sensing image, species tracking with satellites, analysis and early warning of monitoring information, and management activity command.

Information center of the reserve. A conference room and display screen have been built, and the information room of the reserve has been set up for real-time processing of sensing data to achieve intelligent management of the wetland ecosystem of Chenhu Lake.

4.8 Precise layout of innovative beauty and three-in-one aggregation benefits- wetlands in Zhangjia Lake of Tianmen

Zhangjia Lake wetland park enjoys a beautiful natural environment, an important ecological location, a deep cultural heritage and rich wetland resources. The park is rich in rivers and lakes, undulating flat hills, and wetland products, and is known as "the land of fish and rice, the land of sacred water". Zhangjia Lake is a national promotion site for the conservation and sustainable use of *Nymphoides hydrophylla*, and its ecological planting base is famous for its products of "eight immortals in water" (i.e. *Trapa incisa, Sagittaria sagittifolia, Eleocharis dulcis, Nymphoides hydrophylla, Euryale ferox, Calla pallustris, Oenanthe javanica, Zizania latifolia* and other aquatic vegetables). It is one of the three major aquatic reserves of germplasm resources in Tianmen and a protected area for four major fish species including *Mylopharyngodon piceus, Ctenopharyngodon idella, Hypophthalmichthys molitrix, Parabramis pekinensis, Channa argus, Eriocheir sinensis, Cyprinus carpio* and *Carassius auratus*, which are recognized as aquatic green food by the Ministry of Agriculture.

The municipal government will continue to promote the infrastructure construction of Zhangjia Lake wetland park and eco-industrial development as a key construction project of ecological civilization in the city. Based on guaranteeing the benign improvement of ecological functions of Zhangjia Lake, and following the construction policy of "comprehensive conservation, scientific restoration, rational utilization and sustainable development" of the national park, the municipal government carries out the precise layout of the overall construction of the "three-in-one" project, highlighting the characteristics of cultural and leisure experience. In addition, the municipal government insists on taking local resources and innovation as the breakthrough, integrating internal and external resource links and complementing each other, focusing on the three core construction such as scientific conservation and restoration management of wetland ecosystems, scientific research and monitoring and science popularization and education, and infrastructure of wetland industry to build an important demonstration project of China's plain wetland ecosystem, an important educational demonstration base of wetland ecological conservation, and a wetland industry base with Zhangjia Lake wetland park as the core.

4.8.1 Innovate management system, build a support force of comprehensive conservation

The organization and management support system of the city has been built, and sound management mechanism to ensure high-quality development should be guaranteed. In order to strengthen the habitat conservation of the wetland park, the Municipal Party Committee and the Municipal Government set up a leading group of national wetland park conservation, with the secretary of the Municipal Party Committee as commissar and the mayor as team leader, under which special teams are set up to ban poultry farms, comprehensively rectify rural household waste, treat rural non-point source pollution, implement "toilet revolution" and "clear four disorders" of rivers and lakes and river and lake chief patrol. The government implements division of labor, unified action, strict law enforcement and normal management, vigorously promotes ecological, strict and detailed, and efficient management of conservation and construction of the national wetland park. At the same time, the Administration of Hubei Tianmen Zhangjia Lake National Wetland Park and Tianmen Zhangjia Lake Ecological Resources Conservation and Development Co. have been established to assume the functions of conservation, construction, and operation of the wetland park.

4.8.2 Create local characteristics of science popularization, attract public awareness

Establish a system and base for humanistic science popularization and education. With its focus on the construction of the atmosphere of the park as base for humanistic science popularization and education and the centralized effect, the park has added a wetland science and education museum, a wetland nature school, a promenade of science popularization and education, a signage, and a wetland plaza in the first phase of construction. The national park takes advantage of the promenade with a distinctive theme, wetland signage, and guiding signs of tourist traffic to educate and entertain the visitors, citizens, and students to understand wetland knowledge such as the humanistic characteristics of wetlands in Zhangjia Lake, wetland encyclopedia and native plant and animal resources.

Let the activities of science popularization of humanities into the public. The wetland park insists on using the "World Wetlands Day" and "Bird Loving Week" to carry out thematic and informative public welfare activities, promote wetland science knowledge into the authorities, schools, enterprises, communities, families, popularizing the knowledge of wetlands, enhancing the public awareness of wetland conservation, promoting the ecological culture and harmonious coexistence of man and nature. It has been named by the Municipal Science Association as the Science Education Base of Tianmen City.

Create an education base for ecological civilization. The wetland park makes full use of the science and education center, wetland classroom, cognitive garden of wetland plants, bird watching tower and other educational facilities for the public to carry out ecological study of wetlands, nature education activities, making it a newly promoted "national" popular tourist attraction and an education base for public ecological civilization in Tianmen.

Guide social forces to organize different forms of activities of science popularization of humanities. The wetland park sets up a volunteer task force of residents from 9 villages and communities around the wetland park to participate in the wetland conservation and work together to build a beautiful home; makes full use of the professional advantages of social forces such as the City Photographers Association, the Association of Senior Citizens Calligraphers and Painters, and the Society of Poetry and Couplets and provides them with a creative base platform to widely cultivate teams for science popularization.

Integrate media to form a wide impact of communication. The wetland park creates its official accouont of WeChat and Tiktok to timely release construction achievements of the wetland park, beautiful scenery, conservation measures and popular science knowledge to the public. By fully using the Internet, newspapers, television and other media, it carries out popular science propaganda, widely publicizes the knowledge of wetland conservation, and vigorously advocates the concept of ecological civilization.

4.8.3 Create a brand of wetland economy, build a base of ecological industry

Explore the model of co-construction and co-management. The wetland park gives full consideration to the participation of the surrounding residents to drive the development of the surrounding industries and carries out in-depth community building activities. It has held more than twenty seminars, signed a co-construction and co-management agreement with the surrounding villages and established a daily communication mechanism. It regularly organizes villagers to visit and study to enhance the awareness of wetland conservation.

Create a base of wetland industry. The wetland park strives to build a 10,000-mu special industrial base of "eight immortals in water" in the surrounding nine villages, and to construct the operation model of "base + farmers + companies (cooperatives) + e-commerce platform", making every effort to build the e-business industrial park of "eight immortals in water", and promote the sustainable and healthy development of rural economy.

4.9 Carry forward the wisdom of ecological polder to recreate dream water town of Zhuhu in Xiaogan

The Zhuhu wetland park in Xiaogan, Hubei is located in Xiaonan district, Xiaogan city, featuring river wetlands and reservoir and pond wetlands, mainly consisting of three major types: river wetlands, marsh wetlands, and artificial wetlands, and covering five wetland types: permanent rivers, marshes, reservoirs and ponds, water transfer rivers, rice paddies. In December 2018, it was approved by the National Forestry and Grassland Administration as a national wetland park, rated as a national 3A tourist attraction, named as a science education base in Hubei Province by Hubei Provincial Science Association. In 2019 it was approved as an important wetland in Hubei Province; in October 2021 it was included in demonstration pilot project of near-zero carbon emission. It is a photography base in Hubei, and the designated ecological tourism destination for many important tourism companies in the province.

4.9.1 Path with Zhuhu characteristics of returning land for farming to lakes and ecological restoration

After Zhuhu park has been returning land for farming to wetlands, a scene of "people retreating and lakes advancing, and people protecting the lake" has appeared.

Firstly, the government takes comprehensive environmental treatment measures at the level of Zhuhu basin, such as the flood control and drainage of the Fulun River, the reconstruction of post-disaster water conservancy facilities, the raising and thickening of embankment of the Fuhe River, the comprehensive treatment of water environment of the Huanshui River, the demolition and construction of ecological conservation zone, the dredging of channels, the connection of internal canal network and other treatment projects of wetland conservation and ecological restoration. Then, it takes the path with Zhuhu characteristics of ecological restoration: returning land for farming to wetlands and returning farmlands to the lake. It sets up the multifunctional polder field of Xiangyangyuan, cultivating more than 500 edible aquatic plant fields of wetlands, ornamental plant fields of wetlands, medicinal plant fields of wetlands, weaving material plant fields of wetlands, seed plant fields of wetlands. They are of different sizes and forms to enrich biodiversity and promote soil and water coservation, water purification, biological conservation, flood storage and regulation, and micro-climate regulation. It sets up the base/multi-pond system, the forest-marsh compound wetland project combining trees and shrubs, forest belt and forest garden around the park, and plants a large number of native plants in the original farmland system along the lake according to the topography to enrich biodiversity, make full use of space, attract more biological habitats, and restore ecological chains.

The park has replanted more than 30 species of native wild plants such as *Phragmites australis*, *Acorus calamus*, *Typha angustifolia*, *Scirpus validus Vahl*, and *Nelumbo nucifera*, established more than 500 micro wetland groups of 780 acres in its reconstruction areas and conservation and restoration areas, giving full play to the five functions of ecology conservation, water purification, species breeding, creating habitats and beautifying the environment, forming a good state where small wetlands situate in large wetlands and small habitats enhance big environment, creating a unique ecological environment of a harmonious, mutually-beneficial and symbiotic nature, and acheving special results in ecological conservation and restoration. More than 10 kinds of wild plants such as *Euryale ferox*, *Scirpus validus Vahl* and *Brasenia schreberi*, which once disappeared in the Zhuhu Lake area, and more than

20 kinds of wild birds such as *Aythya baeri*, *Cygnus columbianus*, *Anser albifrons* and *Ardea cinerea*, which have not been seen for decades, have reappeared in the rivers, lakes and branches, which are full of vitality. The ecological beauty of "Just heard the birds singing in the east field, then saw the wild lotus lying in the lake." is heartwarming.

After Zhuhu wetland park was included in the demonstration pilot project of near-zero carbon emission of the province in 2021, an ecological community consisting of 30,000 mu of planting base producing national geographical indication protected product "Zhuhu glutinous rice", 20,000 mu of demonstration field of ecological aquatic health breeding, and 20,000 mu of industrial base of high-quality and efficient flower and seedling around the wetland park.

4.9.2 Wetland polder of Zhuhu Lake and eco-industry of lake township

The polder is a kind of farmland developed in ancient times by the working people of China in the process of siltation of the riverside beaches and lakes. People built dikes around low-lying areas, set up culverts on the dikes, closed the gates during normal times to prevent water, and opened the gates to release water into the polder during droughts, so that they could guarantee the harvest whether there was drought or floods. Polder, once played an important role in the development of the Zhuhu Lake, showing scenes of "fields and ditches stretching together" and "paths between the fields look like beautiful embroidery in a staggered pattern". How to inherit the polder culture, so that today's new ecological polder in Zhuhu wetland conservation and ecological agricultural development can assume its rightful historical mission?

Amplify the advantages of wetland features. The wetland park should restore the traditional polder, create ecological polder, give full play to ecological polder with functions of ecological products output, environmental optimization, water purification, biological conservation, rainwater storage and regulation in order to create a wetland ecological conservation model with Zhuhu characteristics.

Amplify the rational use of wetland functions. In its ecological polder area, we have designed more than 500 edible aquatic plant fields of wetlands, ornamental plant fields of wetlands, medicinal plant fields of wetlands and weaving material plant fields of wetlands, to enrich biodiversity, water and soil conservation, water purification, microclimate regulation, and achieve coexistence in synergy between huamn and wetlands.

Amplify the humanistic landscape of the wetlands. Taking into consideration of the historical elements of ancient Yunmeng Marsh, we transformed and restored the original rectangular farmland into the ancient form of polder, which provides an ideal habitat for aquatic plants, benthic organisms, fish and water birds, and also creates a humanistic landscape of wetlands with strong ornamental qualities. In recent years, the park has attracted more than 110 batches of writers, poets, literary enthusiasts, as well as journalists, new media creators to the wetlands of the Zhuhu Lake for sightseeing and folk art collection, and published more than 500 pieces of wetland works in various genres and more than 1,000 pieces of photography in national, provincial and urban media and newspapers at all levels.

Amplify the natural landscape of the wetlands. Through the development and careful cultivation of ecological polder in the wetlands, we broaden the construction horizon and the amout of polder plants has reached more than 200,000 plants of 30 species. At present, the "five areas" of the wetlands and 50,000 mu of rice paddies form together a huge ecological cycle system, and the ecological environment of wetlands basically has restored to the level of that in the early 1970s.

As for the development potential and location advantages of the Zhuhu Lake, we position the direction of economic development of the wetlands in the cultivation of famous products, exploration of special economy, establishment of industrial base, conservation of ecological environment, promotion of integration of agriculture and tourism, focus of adjusting industrial structure, and optimization of regional economic layout, in order to form an industrial park integrating agriculture and tourism, sightseeing and leisure with green four seasons, fragrant glutinous rice, clear water and swimming fish. "glutinous rice of Zhuhu" is a special product of the Zhuhu Lake and a very important part of its wetland ecosystem. The Zhuhu Lake and its surrounding areas have more than 50,000 mu of rice paddies with a humid climate all year round, beautiful water and fragrant rice, which is a resort for *Egretta garzetta, Anas platyrhynchos* and *Anser cygnoides*, and has become an enchanting and idyllic scenery for sightseeing with its high-quality wetland functions and high-efficiency economic functions. The wetland park is a fertile ground for nurturing and inheriting the culture of the lake township, and it is one of the birthplaces of traditional farming food production such as Xiaogan rice wine, sesame candy, glutinous rice cake and tofu skin. During the construction of the wetland park, the private owners were strongly supported to invest more than 30 million yuan to establish Xiaogan Tianlong Wine Co. Ltd. specializing in the research and development and production of series products of "Xiaogan rice wine". With the quality glutinous rice and water of the wetlands, it has developed 16 kinds of new taste "Xiaogan rice wine" and "pearl glutinous wine" with soft package and leisure style, allowing the "Xiaogan rice wine" to go from here to the whole country, to the United States, Singapore, Kazakhstan and many other countries in the world.

4.10 Integrated management of all-for-one wetland conservation and rural revitalization area of Juhe River in Yuan'an

Juhe River wetland park in Yuan'an Hubei follows the policy of "comprehensive conservation, scientific restoration, rational utilization and sustainable development", actively explores the transformation path of "lucid water and lush mountains are invaluable assets", improves wetland biodiversity and stability of composite ecosystem, and heightens the awareness of the whole society on ecological environment conservation in order to provide important guarantee for the construction of ecological civilization and beautiful Yuan'an.

4.10.1 Wetlands of Juhe River in Yuan'an and characteristics of resources

Juhe River wetland park in Yuan'an Hubei is located in the middle reaches of Juhe River and the south-central part of the county, close to Yuan'an County, including Juhe River, the Mingfeng River, channels of the Juzi Creek in the region and the mountains, woodlands and part of the farmland on both banks, with a total length of 59.05 kilometers of the shoreline, which is an important part of the wetland ecosystem of Juhe River Basin and has certain typicality and representativeness in the western part of Hubei and even in the country. The wetlands of Juhe River in Yuan'an have been included in the *List of the First Batch of Wetlands of Provincial Importance in Hubei Province* in 2019 and in the *List of Wetlands of National Importance* in 2020.

The wetlands of Juhe River in Yuan'an are rich in species resources. There are 448 species of plants including 8 species of national key protected plants, 257 species of vertebrate animals, 173 species of birds including 22 species of national key protected animals and *Ciconia nigra*, *Mergus squamatus*, *Ciconia boyciana* and *Syrmaticus reevesii* as the national first-grade key protected ones, and there are 43 species of fish including 13 species of Chinese endemic fish. There are 44 species of provincial key protected animals, 166 species as national protected "three haves", 8 kinds of amphibians distributed in the wetlands listed into the "three haves".

4.10.2 Wetland conservation and management innovation for the whole area

Wetland conservation for the whole area is managed in a coordinated manner. Yuan'an county has established a comprehensive and coordinated mechanism to plan the whole situation and manage the whole area. It has built a county-wide working pattern of wetland park construction and wetland conservation and management, attached importance to ecological governance and wetland conservation and restoration, firmly established the concept of ecology for the whole area, expanded wetland conservation from local pilot to the whole area, and deeply integrated wetland conservation with tourism for the whole area, rural revitalization, urban and rural construction and water conservancy, and transportation infrastructure; it clearly proposed ideas for wetland conservation and construction of "ecological priority, comprehensive conservation, respect for nature, reflecting nostalgia, comprehensive governance, low-carbon pollution control, highlighting the function, scientific restoration".

Weaving wetland conservation network in all directions. Yuan'an county has included conservation areas and restoration and reconstruction areas of the wetland park into the ecological redline control, deployed a full range of conservation forces, and constructed a network of inspection and conservation of "caretaker + river chief + volunteer" of the wetland park. Patrol and supervision is carried out in the way of "all staff to the front line" by Management Office of Wetland Park and Yuan'an County Forestry Administration. Yuan'an county strongly promotes joint supervision and carries out special operations of "clearing four chaos", "action to welcome spring" and "action of clear flow" with County Branch of Ecological Environment Bureau, Water Conservancy Bureau, and Bureau of Agricultural and Rural Affairs to implement ecological and environmental improvement of Juhe River basin. The establishment of wildlife habitat conservation points in Pipazhou and Chenjiapeng, video monitoring platform of the wetland park and monitoring points as well as the standardization and improvement of the management facilities have basically covered the online monitoring of the park river and imporved the conservation and management level of the park.

4.10.3 Promoting conservation of small and micro wetlands to integrate rural revitalization and all-for-one tourism

The county firmly establishes the concept of harmonious coexistence between human and nature, places construction of ecological civilization in a more prominent position, pragmatically does a good job in "three articles" of ecological restoration, environmental conservation and green development, and strives to make the favorable ecological environment become an important support point and growth point for high-quality development in Yuan'an. The county adheres to the policy of giving priority to conservation and protection and natural restoration as the main part, ponders the important connotation expressed by wetlands being called "the kidney of the earth", elevates the design concept of wetland park to the height of the fundamental guidelines for the implementation of the strategy of rural revitalization, promotes the construction of small and micro wetlands to be fully integrated into the construction of green towns and the development of tourism for the whole area, and enhances the construction and conservation of small and micro wetlands scientifically and make efforts to guard the ecological background by enhancing ecological functions and ecological values and making reasonable use of native natural resources.

Carry out county-wide conservation model for small and micro wetlands. The county carries out ecological restoration through a combination of natural and artificial ways to protect wetland resources and restore ecological functions. It has successively built small and micro wetlands in the old county town of Dongjia section of the Luyuan River, Bamangdian Village of the Wuli River in Yangping Town, West Lake weir in Mingfeng Town, Madao weir, Mingfeng estuary, and the north section of the 3rd bridge of Juhe River where the land for farming has been returned to wetlands.

Promote utilization model of special small and micro wetlands. The county combines conservation of wetland resources and promotion of agricultural development, and constructs production-oriented small and micro wetlands. It encourages and guides farmers and rural business entities to develop water-related planting and breeding with characteristics, promotes the construction and conservation of small and micro wetlands, advances the development of leisure agriculture and rural tourism, to achieve the dual enhancement of economic value and ecological value. It has successively built rural small and micro wetlands in Jinjiawan, ecological wetland scenic areas such as Taohua Island Ecological Park, as well as courtyard small and micro wetlands in Jiuzi Creek community in Mingfeng Town, Shuangquan community, and Hongfu estate in the old town.

Explore co-building model of wetland popularization and education and community. The park focuses closely on the strategic layout of "rural revitalization" and strives to explore the model of community co-construction. The park has built wetland nature schools, wetland nature libraries, and popularization and education galleries of wetland plants, integrating the concept of ecological civilization into the daily life of the public, so that the public can experience a strong ecological cultural atmosphere in leisure and fitness.

4.11 Practicing the theory of two mountains, exploring the path of quality green development of three-life integration in Xujiahe River

4.11.1 Characteristics of wetland functions in Xujiahe River and water source conservation

Xujiahe National Wetland Park is built on the basis of Xujiahe reservoir, which belongs to the Fu River water system. The main rivers upstream include four major tributaries such as Xujiahe River and the Longquan River, all of which originate from the southern foot of the Tongbai Mountain and converge into Xujiahe reservoir. Xujiahe reservoir is the third largest reservoir in Hubei Province, and its original water recharge is mainly based on the comprehensive recharge of atmospheric precipitation and surface runoff, now supplemented by the water resource allocation project of Danjiangkou recharge in northern Hubei. The total capacity of the reservoir is 778 million cubic meters, with 299 million cubic meters of storage capacity of reservoir and a multi-year average inflow of 6.7 cubic meters/sec. The reservoir basin is full of low hills, islands and branches, and is a typical reservoir type of hilly branching. Xujiahe reservoir is called the Pearl of Northern Hubei, which combines ecological functions such as water source, ecology, flood control, flood storage, and tourism.

Xujiahe National Wetland Park has a composite wetland ecosystem type with biodiversity, wetland landscape of reservoirs and human living environment, and its multiple wetland complexes form a unique wetland ecosystem. Xujiahe reservoir has a vast water area, rich wetland wildlife resources, remarkable features of reservoir wetlands, natural wetland morphology, beautiful landscape and strong ornamental value, and is an important part of the wetlands in the middle and lower reaches of the Yangtze River with strong representativeness.

Xujiahe reservoir is an important water source, assuming an important regional function of drinking water source and of irrigation water source. The government has formulated the Comprehensive Treatment and Implementation Plan of Xujiahe Reservoir in Guangshui to take comprehensive treatment action for wetland conservation of Xujiahe River, and a series of environment treatment and ecological governance in and around the Xujiahe Wetland Park continuously, which has achieved results and significantly improved the water ecology and water environment of Xujiahe River. First, the government has cancelled cage farming, removing and outlawing cages and block nets om the Xujiahe reservoir and clearing farming on water surface. Second, the government has rectified poultry farming by investigating and handling all kinds of farms (breeders) and controlling the source of pollution. Third, the government has prohibited river sand mining. It issued a city-wide notice on combating illegal sand mining in the reservoir area of Xujiahe River to strictly implement the river chief scheme and departmental responsibilities, and crack down on illegal sand mining in the reservoir area. Fourth, the government has rectified illegal construction by demolishing the illegal construction of villas and other illegal buildings, reclaiming and planting greenery, eliminating pollution hazards and maintaining the ecological integrity of the shoreline. Fifth, the government has launched the ban on fishing in the Xujiahe reservoir. Sixth, the government has increased fish stocking by releasing fish fry of *Hypophthalmichthys molitrix*, *Hypophthalmichthys nobilis*, *Ctenopharyngodon idella*, and *Parabramis pekinensis* to enrich the biological resources of Xujiahe River. Seventh, the government has carried out the project of domestic sewage treatment into the reservoir. Eighth, the government has removed illegal dwarf fences, which greatly ensures the smooth flow of the water system and water connectivity, prevents pollution from breeding and improves the ecological environment of wetlands.

At present, the irrigation zone of Xujiahe River involves five counties and cities of Guangshui, Anlu, Yunmeng, Xiaonan and Xiaochang, with an irrigated farmland area of 38,000 hectares. It is a large type II reservoir mainly for flood control and irrigation, with a comprehensive use of water supply, shipping and breeding. It is also a source of drinking water and domestic water for more than one million people in Guangshui, Anlu, Yunmeng, Xiaochang and Zengdu.

4.11.2 Model of community co-construction under the concept of integrated development of three lives

Create beautiful villages around the wetlands and highlight the characteristics of the creation. The government connects the construction of a wetland park and the construction of a beautiful countryside in an organic manner, and steadily promotes the co-construction of wetland parks and communities, having further created a number of co-construction sites of wetlands and communities such as Yinghua Shizi, Huayun Quanshui and Renwen Xiaodian, highlighting the characteristics of the creation, and allowing wetland conservation and construction and production and life of the surrounding residents to complement one another well.

(1) co-construction site in the Shizigang village in Maping town. The Shizigang village is located in the eastern part of Maping town, the west bank of the Xujiahe National Wetland Park. It has won the honorary titles of National Forest Village, Provincial Green Model Village, Provincial Migrant Beautiful Home and Provincial Health Village. With the Yinghua Garden covering an area of more than 1,000 mu it has built, the Shizigang village now enjoys a good ecological environment and ecological tourism value. The wetland park and the village have co-built a joint management site, which will further strengthen the conservation force of the wetlands in Xujiahe River and enhance the ecological environment of the wetlands.

(2) co-construction site in the Meimiao village in Guanmiao town. The Meimiao village is located in the southeast of Guanmiao town, the north bank of the Xujiahe National Wetland Park. It has won the honorary titles of National Model Village of Rural Governance, National Forest Village, Ecological Tourism Village in Hubei Province. With a Gesang Garden covering an area of more than 1,000 mu, a Yueji Garden covering more than 500 mu, and other seedling gardens covering 1,000 mu it has built, the Meimiao village enjoys a beautiful ecological environment and the basis of the management team. A joint management site to be built with the village will enhance the management capacity of the wetland park and expand the influence of wetland science and education with the least investment.

(3) the Quanshui village in Changling town. The Quanshui village is located in the

northeast of Changling town, Guangshui city, the eastern bank of the Xujiahe National Wetland Park. It has been awarded as one of the AAA-level tourist attractions, won the honorary titles of Provincial Green Model Village and Top Ten Famous Tourist Villages, and nominated Top Ten Most Beautiful Countryside in Jingchu. It develops industries of flower and seedling planting and ecological breeding. Taking "beautiful countryside with flowers and spring" as its core, the village has established itself a good ecological tourist attraction. A joint management site to be built with the village will further enhance the ecological tourism image of the Xujiahe National Wetland Park and lay the foundation for the later development of tourism.

(4) the Xin'an village in Changling town (co-construction site of Mingyue Bay Resort). The Xin'an village in Changling town (Mingyue Bay Resort) is located in the middle of Changling town, Guangshui city, the south bank of the Xujiahe National Wetland Park. The Mingyue Bay Resort has been awarded as one of the AAA-level tourist attractions, and is the site for Creation Base of Guangshui Writers' Association, Creation Base for Guangshui Poetry, Couplet and Verse, and Tour Base for Suizhou Study. It has won the honorary titles of National Five-Star Agritainment by Hubei Provincial Department of Culture and Tourism and Demonstration Point of Leisure and Agriculture in Hubei Province. It is a rural ecological tourism destination invested in 2004 by Guangshui City Xujiahe Shuihu Culture Tourism Co. Ltd. with the core of culture of the poet Li Bai and combination of leisure tourism, agricultural science and study tourism. The Mingyue Bay Resort is now the most influential ecological tourist attraction around the Xujiahe wetlands, with a good ecological environment and management basis. A joint management site to be built with the village will further strengthen the ecological conservation of the wetlands and publicity efforts.

4.11.3 Towards comprehensive wetland conservation and rural revitalization with high quality

Guangshui city explores and practices the "two-mountain theory" and creates a demonstration area for high-quality integrated development. The city has proposed the goal of "building a demonstration area for high-quality integrated development of Yueguanghai", combining with the construction of the creation and management of the Xujiahe National Wetland Park to formulate an action plan, meet the basic requirements of accelerating the establishment of the "two-mountain theory" practice area, demonstration area of rural revitalization, benchmark area of ecological conservation, model area of industrial integration, and model area of grassroots governance, and ultimately realize an important guarantee for the common prosperity of the people in the region.

Leading by planning in a scientific manner. Based on the principles of putting conservation at the first place, giving priority to ecology, leading by planning, linking the whole area, implementing step by step and focusing on building, the government firmly establishes the concept of "two mountains", confirms the national tourist resort and national wetland park as the starting points, and takes the integrated development path of building "six districts and one park". Through the introduction of a high-level planning team, the government has created action plans and implementation plans for creating model areas, launched planning and the preparation of the starting project, to steadily promote the ecological conservation and construction of wetland parks, and make a good start for high-quality integrated development.

Sorting out policies and resources. First, the policy requirements. They include the provisions and requirements of the Wetland Conservation Act, the requirements for the conservation of drinking water sources, the scope of first-level and second-level conservation, the scope and functional zoning requirements of the wetland parks, the scope and requirements of reservoir management, and the scope and requirements of the conservation site of Salangidae germplasm resource. Second, the cultural heritage. It mainly includes Shoushan culture of Li Bai, food culture of King Zhan, fishermen culture of reservoir, and red culture of the Yuntai Mountain. Third, the stock resources. We have inventoried in a comprehensive and detailed manner of seven idle and available resources such as traffic training center, finance school, personnel hall, sanatorium, Liujiao island, Taohua island and Longfeng island to ensure one ledger for one place. Fourth, the industries with characteristic. The demonstration areas focus on cultivating industries such as tea, oil tea, vegetables, fragrant rice, rape, Salvia chinensis Benth, and flowers and seedlings.

Strengthening conservation and control of natural resources in demonstration areas. Notice on Strengthening Conservation and Utilization, Control of Planning and Construction of Natural Resources in Demonstration Areas has been formulated and issued to the relevant member units and five town offices in the name of the command.

4.12 Returning land for levees to lakes to restore the natural ecological beauty of Diaocha Lake in Hanchuan

4.12.1 Evolution of Diaocha Lake and its current situation of wetlands

Diaocha Lake is located in the lower reaches of the Hanjiang River, the northern edge of the Jianghan Plain, and the northern part of Hanchuan city, Hubei Province, and is an artificial closed lake formed by four trunk canals from four directions. The exterior of the lake is strictly controlled by the four canals and is rectangular in shape with a length of 16.1 kilometers from the east to the west and a width of 5.5 kilometers from the north to the south. Diaocha Lake was originally a floodplain, a lake where the water came to be a vast ocean and the water receded to be a desolate lake. Before the 1950s, many lakes of all sizes here formed a lake group and many small and medium-sized rivers such as the original Tianmen River, the Ai River, the Silong River, the Dafu River and the Fuhe River injected into the Hanjiang River through Diaocha Lake.

From 1969 to 1972, Diaocha Lake was reclaimed as the largest inland closed freshwater lake in China with the main function to drain floods and reduce peaks to ensure the safety of people's properties in the basin. During that time, Diaocha Lake was developed in the way of "storage and regulation as the focus combined with aquaculture", and nearly 100 square kilometers of the Lake was enclosed into a scattering of intensive fish ponds of different sizes, leading to the lake surface's shrinking faster.

In 2014, the National Forestry Administration approved Diaocha Lake as a pilot construction of national wetland park. in 2017, the Diaoyu Lake was listed as a unit of river and lake chief scheme in Hubei Province, and in 2018, the Diaoyu Lake was listed as a key treatment project of the well-coordinated environmental conservation of the Yangtze River. It is currently aimed at restoring and protecting the original wetland ecosystem of Diaocha Lake and making reasonable use of wetland resources. Diaocha Lake ushers in the historical opportunity of "six songs of lakes (forming the lake by returning land for levees, seeking orders of the lake by rule of law, beautifying the lake with ecological environment, bringing prosperity to the lake through industrial development, making the lake special with its characteristics, underpinning the lake by innovation)".

4.12.2 Returning land for levees (farming and fishing) to lakes and ecological restoration of Diaocha Lake

In March 2019, the People's Government of Hubei Province approved the conservation plans of ten transboundary lakes, and Diaocha Lake in Hanchuan was included. The implementation of the Plan will effectively guarantee the stability of the lake form; the water surface (volume) will not be reduced; the development and utilization will be effectively controlled; the ecology of the lake will achieve a virtuous cycle; the comprehensive functions of the lake will be given a full play; and the ecological conservation of the lake and the sustainable use of resources will be realized. The "three returning" of Diaocha Lake in Hanchuan, i.e. the construction of the national wetland park is the strategic decision of Hanchuan municipal government to strive for a green development of ecology for the benefit of its future generations.

First, the government carries out the project of returning land for levees (fishing and farming) to wetlands. The implementation scope of the project of returning land for levees (farming and fishing) to lakes (three returning) of Diaocha Lake is the flood storage area (farming area) of 48.7 square kilometers, involving about 2,189 households of six municipal units, seven administrative villages and nine state teams in the farming area of Diaocha Lake with a total compensation fund of 1.178 billion yuan issued. In addition, new communities are built to resettle landed fishermen; pensions are issued to solve the aging problem of fishermen; and entrepreneurial training is carried out and employment positions are provided to help landed fishermen reemploy.

Second, the government carries out ecological treatment and ecological restoration. The ecological treatment mainly includes connection of water systems, dredging, and raising and thickening of side dikes. The connection of water systems mainly means intermittently removing the embankments of the original farming ponds in the restoration and reconstruction area, so that the water bodies in the lake can be connected to each other. The length of removed embankments totals 18.5km with the earthwork of 3.471 million m^3. The ecological restoration projects include restoration of terrestrial vegetation along the lake, restoration of aquatic vegetation in the lake, plant filtering zone, shoal and bird island, and habitat restoration.

4.12.3 Ecological restoration and treatment effectiveness of Diaocha Lake

More than 60,000 mu of embankments of Diaocha Lake were demolished and returned to the lake, fully realizing "three returning" of 48.7 square kilometers of fish ponds in the storage area. One million fish fry have beem released; 660,000 square meters of *Scirpoides holoschoenus* have been planted; the lake surface has been enlarged about 20 times compared with that before "three returning"; the ecology and related functions of the lake have been greatly improved. The flood storage capacity of Diaocha Lake has been greatly enhanced, reaching 120 million cubic meters.

At present, by carrying out a series of wetland conservation and restoration work such as returning land for levees (farming and ponds) to wetlands, removing seine-nets, connecting water systems, dredging, and restoring aquatic vegetation, the ecological environment of Diaocha Lake has continued to improve, creating favorable conditions for wildlife to flourish. The water quality of Diaocha Lake has risen from standard V to standard III; the species and number of wild plants and animals in the area have increased significantly. There are 67 new species of vascular plants and 9 new species of birds, and the number of national protected animals such as *Cygnus columbianus* and *Platalea leucorodia* has increased more than ten times.

In order to fully promote the "three returning" of Diaocha Lake, the government actively explores the quality green development path of "wetlands + park", "ecology + tourism" and "entrepreneurship + employment", and fully relys on the characteristics of the natural resources of Diaocha Lake basin to protect the lakes, forests, farmlands and grasslands of the Diaocha Lake, fish, shrimp, animals, insects and birds of the wetlands with the help of the national strategy of all-for-one tourism. At the same time, the government combines ecological culture, water conservancy culture, fishery culture , historical culture and popular science education to create a unique national artificial inland "first lake", and actively solves the production and living problems of the people in the lake area by determining the resettlement compensation program to solve their problems of livelihood.

4.13 New practices of wetland conservation and regional development in green Chibi

4.13.1 Typical lake wetlands in Chibi city area

Chibi, located on the south bank of the middle reaches of the Yangtze River, has a deep historical and cultural background and rich geographical and natural resources. The wetlands in Chibi, represented by Lushui Lake, Huanggai Lake and Xiliang Lake, have made extremely excellent and unique wetland ecosystem.

(1) Lushui Lake. The water area is 57 square kilometers, with 720 million cubic meters of water storage. More than 800 islands are scattered in the lake. The main dam of the lake is located in the upper reaches of the land water of Chibi city, which is a test dam of the Three Gorges Project. The No. 8 secondary dam is the longest artificial clay dam in Asia, and the Lushui River below the main dam meanders through the green hills and fields of Chibi for 40 kilometers before injecting into the Yangtze River from the town of Chibi, the site of the Battle of Chibi of the Three Kingdoms. The Lushui Lake has excellent water quality and rich aquatic resources, integrating water storage, flood regulation, power generation and breeding; the lake area has a wide variety of flora and fauna, green mountains and beautiful water, and a pleasant scenery.

(2) Huanggai Lake. Located at the junction of Hunan and Hubei provinces, the Huanggai Lake is connected to the Yangtze River in the north and Honghu Lake in the west, and is a natural lake shared by Hunan and Hubei provinces. Historically, the Huanggai Lake was once not well known in ancient Yunmeng Marsh. The Battle of Chibi at the end of the Eastern Han Dynasty not only made the rock spur of Chibi famous in the world, but also gave this thousand-year-old puddle on ancient Yunmeng Marsh a name of its own - Huanggai Lake. The unique geographical location and ecological environment of wetlands in the Huanggai Lake has formed a rich wetland flora and fauna resources, and there are many rare and endangered species, presenting features of high proportion of rare species and large size of their popula-

tion. The vast water and wide wetlands in the area are an important transit point and habitat for birds' wintering and migration. At the same time, the Huanggai Lake is a river-connected lake, connecting to the Yangtze River for nine months of the year. The species resources can freely enter, exit and exchange, forming rich biological resources of the wetlands and providing convenient conditions for species diversity.

(3) Xiliang Lake. Located in the northeast corner of Chibi city, the Xiliang Lake has a low topography, a dense water network and connected rivers and streams. Its vast 82-square-kilometer water (the towns along the lake in Chibi city account for about 60% of the total water area of the lake, while the towns along the lake in Jiayu county and Xian'an district account for about 40% of the total water area of the lake) is covered with smoke and mist. The *Nelumbo nucifera*, *Euryale ferox*, *Lactuca sativa* var. *angustana*, *Trapa natans*, and *Brasenia schreberi* in the lake prosper in all seasons; many other freshwater fish such as *Mylopharyngodon piceus*, *Cyprinus carpio*, *Ctenopharyngodon idella*, *Carassius auratus*, *Parabramis pekinensis*, *Siniperca chuatsi*, *Channa argus*, *Hemiculter leucisculus*, *Tanichthys albonubes*, *Monopterus albus*, *Testudines*, *Trionychidae*, Brachyura and Caridea swim freely in it; it is also a great place for water birds such as *Anas platyrhynchos* and *Egretta garzetta* to find a mate and and a habitat for making a nest.

4.13.2 Practices of wetland conservation and management development in Chibi

As the first batch of national wetland parks built in Hubei Province, the Lushui Lake National Wetland Park has assumed the historical mission of green development from the day it was established. In January 2017, according to the work plan of the Municipal Government, the division of responsibilities of the Management Office of Lushui Lake National Wetland Park was clarified to undertake the conservation and management of the wetlands of the city including that of the Lushui Lake.

As a form of nature reserve, the construction of national wetland parks is carried out according to the principle of one policy for one place. The management model of Lushui Lake National Wetland Park in Chibi has opened a new model and new way of wetland conservation and development in a way.

Avoiding duplication of institutions and intensive use of financial resources, and allowing professionals to do professional work. The practices have proven that the new practices of wetland conservation and development in Chibi have not only effectively protected the wetland resources in Chibi, but more importantly, have effectively dovetailed with the integration and optimization reform of national park-based nature reserves which is being promoted by China, showing strong effectiveness of ecological conservation and regional vitality of green development.

Since 2016, in accordance with the national instructions to "promote well-coordinated environmental conservation and avoiding excessive development" and the "ten-year ban on fishing" of the Yangtze River basin, Chibi city has launched an action of a "ban on fishing" with the main content of removing all "nanzhu (bamboo piles), nets (cages), ground cages and bewitching traps" and "landing fishermen". With the removal of bamboo piles and the full implementation of the "ten-year ban on fishing", the picturesque images of "the sunset and herons fly together, green water and blue sky share one color" of Xiliang Lake and Huanggai Lake, with the clarity and purity of water, are beautifully presented as a charming scenery surrounded by mountains and water with hundreds of turns.

4.13.3 Effectiveness of ecological conservation of wetlands in Chibi

The superior geographical environment and humid climate of the Huanggai Lake have nurtured a rich biodiversity. Among the huge biomass recorded of the Lake, birds almost occupy the most part. Taking wintering migratory birds as an example, the number of water birds that winter in the lake area reaches 40,000-50,000 every year. The turn of winter and spring is the peak time for wintering birds. We can find flocks of birds whether on the wide and endless water, or among the tidal flats of marshes. According to the scientific records of autumn and winter in 2021 alone, the number of birds recorded in the survey has been 67,757, belonging to 15 orders, 42 families and 128 species, of which, 41 species of 7 orders and 11 families of water birds. The survey recorded 7 species of national key protected birds, including Leucogeranus leucogeranus, one species of national Class I protected animal, and a total of six species of national Class II protected animals such as *Cygnus columbianus*, *Milvus migrans*, *Circus cyaneus*, *Falco tinnunculus*, *Falco peregrinus* and *Centropus bengalensis*. The rich water bird resources and the high conservation value of species rarity in the Huanggai Lake have attracted the high attention of the whole society.

4.14 Ecological restoration of urban lakes to create a treatment model of lakes in Hubei

According to the *White Paper on Lake Conservation and Management in Hubei Province in 2021* released by the Hubei Provincial Water Resources Bureau in July 2022, among 755 lakes included in the provincial list of lake conservation, 293 lakes have completed ecological restoration planning, and 120 projects are undergoing ecological restoration (including 86 projects of dredging and comprehensive treatment). In 2021 overall water quality of the major lakes in the province is mildly polluted, with the main pollution indicators of total phosphorus and chemical oxygen demand. Among the 29 waters of the 24 provincially controlled lakes, the water quality of 31.0% of them is II~III, 69.0% IV and V, and no poor V. Among the 29 waters, two of them have medium nutrient status and 27 are slightly eutrophic. Compared with the same period in 2020, among the comparable 18 waters of lakes, two of them have improved water quality on a year-on-year basis, ten remained stable, and six declined. Lake conservation and management in Hubei Province has a long way to go, and there is great potential for ecological restoration work.

In the natural environment of Wuhan, lakes are the landscape with the most characteristics. Known as the "river city", Wuhan lives together with water, and has a reputation of "city of a hundred lakes". Wuhan is beautiful because of the lake, and thrives because of the water. The environmental and ecological problems of the lakes in the city are closely related to the social, economic development and urban expansion of the basin. As a "double pilot" city for national water ecological conservation and restoration as well as for building a water-saving society, Wuhan government attaches great importance to the management of lakes, and with the orderly promotion of river and lake chief scheme of Wuhan, implements a series of prevention and treatment projects of lake pollution, significantly improving the ecological environment of the lake water. A variety of technology systems and model practices of urban lake ecological restoration have been formed in Wuhan.

4.14.1 "Neisha Lake model" of ecological restoration of urban lakes

Wuhan "Neisha Lake model" is a new model to restore degraded ecosystem of urban lakes. It proposes "comprehensive pollution interception as the premise, ecological survey as the basis, construction of submerged vegetation as the focus, control of fish community as the means, adjustment of ecosystem structure as the core, with consideration of the landscape" with combination of Hubei characteristics as "combination of Chinese and Western medicine" to address the current ecological problems of the lake caused by large-scale aquaculture in Hubei, including "degradation and disappearance of submerged vegetation, deep and black sludge with stench, and seriously polluted water body".

The "Neisha Lake model", on the basis of comprehensive and effective pollution interception, significantly improves the redox potential at the sludge-water interface through in-situ ecological bottom improvement, and effectively reduces endogenous pollution; through biological membrane technology and in-situ phosphorus locking technology, it quickly realizes the reduction of water transparency and nutrient concentration, and provides good basic conditions for restoration of submerged vegetation. On top of this, through the step-by-step restoration technology of submerged plant community, the model rapidly constructs submerged vegetation and improves quality of water body. Finally, through the construction of food webs based on material and energy flows, the model effectively improves the species diversity and stability of the ecosystem, rebuilds a healthy ecosystem of water body, restores the self-purification capacity of the lake, realizing the lasting health and stability of the lake ecosystem.

The "Neisha Lake model", focusing on improving the substrate of the lake yet with no dredging, saves investment and produces fast results compared with the traditional dredging treatment. The technology is applied to improving the water quality of the Neisha Lake in Wuhan, which has been stable for nine years to reach the standard IV and above of surface water. At present, the technology has been applied to many lakes in the country for restoration.

The "Neisha Lake model" has also been applied to the Dusi Lake, Lingjiao Lake, Simei Pond, Xiaonan Lake, Xibei Lake, Houxiang River, East Lake, South Lake, South Taizi Lake, Zhangbi Lake, Zhuye Lake, Jin Lake, Shangjin Lake, Moshui Lake, Dugong Lake, as well as Yi'ai Lake in Huanggang and Yinjia Lake in Daye, Huangshi, all of which have achieved remarkable results.

4.14.2 "East Lake model" of ecological restoration of urban lakes

The East Lake in Wuhan is the largest lake in the city and assumes important storage function with long fetch, high wind and wave impact, complex surrounding environmental conditions, and. It is difficult to carry out ecological restoration directly in this kind of lake.

The model takes advantage of ecological enclosure to first realize the ecological restoration in part of the area through the construction of a demonstration area of ecological restoration in part of the East Lake bay, and then build a pioneering area of ecological restoration in the important shoreline such as the Tingtao scenic area. After a connection of the demonstration area and the pioneering area, ecological restoration can be realized in some sub-lakes. Through the combination of "point - line – surface", the East Lake steadily promotes the ecological restoration of each sub-lake according to local conditions, and then realizes the ecological restoration of the whole East Lake through the natural expansion of aquatic vegetation.

At present, this "East Lake model" of ecological restoration of urban lakes has been applied in the South Lake, Tangxun Lake, Dugong Lake and Jinyin Lake in Wuhan, and has achieved satisfactory results.

4.14.3 "Shahu Lake model" for ecological restoration of urban lakes

The "Shahu Lake model" for ecological restoration of urban lakes is developed on the basis of the "East Lake model". The Shahu Lake in Wuhan is a regulation and storage lake with river flowing in and out. The water of the East Lake goes westward through the Shuiguo Lake and Chu River during flood season, enters the Shahu Lake, and finally discharges itself into the river through the pumping station at Xinsheng Road. Due to the difficulty in transformation of rain and sewage diversion in the old city and the long duration, the lake is under impact of a large pollution load combining rain and sewage into the lake in the flood season. The lake has characteristics such as a large fluctuations in water level, water quality and water ecosystem by external impact.

The "Shahu Lake model" takes advantage of ecological enclosure to build a channel to guide the upstream water into the enclosed storage pond from the Shuiguo Lake to the Chu River and the Shahu Lake bridge during the flood season to have the preliminary precipitation, and discharge the water into the river through the pumping station at Xinsheng road, which can keep the ecological restoration area outside the containment stable. After the flood season, through the measures such as in-situ improvement of substrate, the model quickly improves the water quality of the water body in the channel to promote the self-recovery and reconstruction of water ecosystem in the channel.

第五篇

湖北湿地保护典型人物与组织

CHAPTER 5

TYPICAL FIGURES AND ORGANIZATIONS OF WETLAND CONSERVATION IN HUBEI

湖北湿地保护建设的历程在遵循人与自然、社会和谐发展的道路上不断探索推进，一代又一代保护者接力奋进，创新开拓，默守奉献，湖北湿地保护成果丰硕喜人，人与自然和谐共生的美好图景徐徐铺展。

在这个历程中，千千万万默守奉献在湿地保护和管理第一线的人员，成为美丽湖北图景中那道亮丽的风景线。在他们中间，有从事湿地保护的著名科学家、教授，几十年扎实工作、奔走呼吁，推动国家和地方政府的决策；有在保护和管理一线的湿地管理者，把最美的青春年华奉献给了湖北的山川林草湿地；有自然保护地的守护者和志愿者们，把湿地保护作为至高的追求；同时，还有一些社会环保组织积极投入到湖北湿地保护中，把和谐共生的理念普及千家，将成果惠及万户。

一、毕力一生，守护"地球之肾"——著名湿地科学家蔡述明

"研究长江中游湖泊的演化、湿地的保护与利用是我一生的专业。我大学学的就是地质地理系自然地理专业，工作又在中国科学院测量与地球物理研究所。20世纪60年代我就来到湖北工作。湖北是'千湖之省'，武汉又是'百湖之市'，这里离洞庭湖、鄱阳湖都不远，在这里研究湿地有着得天独厚的条件。湿地是大自然赋予人类的财富，很美，在野外工作也心情愉快。一辈子都与它打交道，很幸运。"

——蔡述明

2005年11月8日，在乌干达召开的《湿地公约》第九届缔约方大会上，一位来自中国的科学家获得《湿地公约》最高奖——"拉姆萨尔湿地保护科学奖"，他就是中国著名湿地学家蔡述明。湿地保护科学奖设于1996年，蔡述明是该奖设立以来，我国第一位获此荣誉的科学家。

2002年3月5日，时任国务院总理朱镕基在《政府工作报告》中郑重提出"加强湿地保护"。台下，一位老专家的心情难以平静，他叫蔡述明，全国政协委员、时任湖北省政协副主席、中国科学院研究员，他激动地说："这是第一次，以往历次《政府工作报告》都没有像这一次专门提到'加强湿地保护'，说明国家越来越重视湿地问题"。2000年"两会"召开时，蔡述明委员曾经呼吁国家重视对湿地的保护，2002年"两会"期间他再次提出，国家应尽快制定"湿地保护法"，确保国家经济、社会

蔡述明获得"拉姆萨尔湿地保护科学奖"
Cai Shuming receives the Ramsar Wetland Scientific Award

的可持续发展。

从20世纪60年代大学毕业踏上江汉平原这片土地开始,蔡述明就倾心于湿地:研究湿地、认识湿地、保护湿地。在与湿地打交道的50余载,蔡述明始终充满热情,一点也不觉得湿地研究是件枯燥的事情。"戴着草帽、挂着望远镜,坐在摇晃不定的考察船上,行进在湖泊、芦苇荡、沼泽地里,不论酷暑严寒常年进行湿地野外调查工作。一回到科研单位就扎进实验室,尽快对一手数据进行分析与整理,找出湿地环境变化的规律。"这就是蔡述明的日常工作。

通过在湿地科学领域长期的努力和潜心研究他获得了一系列成就:①系统地研究了长江中游江汉湖群和洞庭湖环境演化;②创新性地提出北亚热带湖泊沼泽化的正逆效应理论,研究结论和有效防治措施已为国家决策部门所采用;③提出了适合长江流域湿地生态特点的农业开发模式,其成果得到推广应用。

1997年,蔡述明任湖北省政协副主席、全国政协委员。从此,他肩负着一个政协委员的使命,而他的事业也踏上了一个新的征程——科学进言、保护湿地。

1998年,他以全国政协委员和科学家的身份多次在长江中游的湖北、湖南、江西三省平原湖区进行实地考察,支持和促进长江中游湿地保护工作的开展。他

民进中央洞庭湖暨长江中游湿地保护与开发研讨会（左二为蔡述明，左三为时任民进中央常务副主席张怀西，2002年）

the Seminar on Wetland Protection and Development in Dongting Lake and the Middle Yangtze River (second from the left is Cai Shuming) in 2002

努力推动江汉平原洪湖国家级自然保护区的立项和建设；同时，对洞庭湖和鄱阳湖的国家级自然保护区建设也提出了积极的意见和建议。

作为全国第九、第十届政协委员，他在历次全国政协会议上多次发表意见和建议：呼吁国家保护湿地，重视湿地的生态环境建设。

2002年，在全国政协九届五次会议上他提出"关于长江中游湿地保护与合理利用"的建议，呼吁要将湿地保护与合理利用纳入法制轨道，并做了"关于国家应尽快制定湿地保护法，确保经济、社会可持续发展的建议"的发言。他的湿地保护建议提案成为后来全国人大湿地立法的重要基础。

同年，他作为中国民主促进会会员，为民进中央起草了有关湿地保护和湿地立法的建议。此项建议得到时任国家主席的采纳。当年国务院总理的政府工作报告中也增加了湿地保护的内容，在国内外引起了很大的反响，对湿地保护起了决定性的作用。

2004年，全国政协第十届二次会议上，为确保"一江清水送北京"，他提出"把南水北调水源地汉江上游丹江口库区建设成为国家级自然保护区"的建议。同年5月，作为全国政协委员，他陪同全国政协常委等就南水北调中线工程水源保障问题开展专题调研,到十堰、丹江口等地考察水土流失情况。当时他提出了六条建议,这六条建议均被全国政协向中共中央、国务院报送的《关于南水北调中线工程水源保障问题调研报告》所采用。

2004年10月,他参加并主持了"长江流域省市政协长江水环境第五次研讨会"。会后,及时向国务院呈报了长江流域13省市政协水环境保护提案,全国政协给予此提案高度评价。

在多次对洪湖深入调查研究的基础上,他总结并编写了《湖北洪湖湿地自然保护区综合科学考察报告》。蔡述明对洪湖湿地生态环境遭到破坏的现状十分担忧,于是向时任中共中央政治局委员、湖北省委书记俞正声同志提交了《关于优化洪湖管理,保护湿地资源》的建议,引起了省委、省政府的高度重视。2004年11月29日,湖北省委、省政府在洪湖召开了"加强洪湖生态保护办公会"。时任湖北省委书记俞正声、省长罗清泉等省委省政府领导及20多位省直委办厅局领导以及荆州市、洪湖市、监利县的领导出席了会议。会上,俞书记做了重要讲话,提出了整治洪湖的指导思想和具体措施。如此高规格的关于湿地生态保护的现场办公会,在湖北省还是第一次,在全国也不多见,引起了强烈的社会反响。

2006年7月,蔡述明及其团队对神农架大九湖开展了综合科学考察,并撰写了《关于神农架大九湖湿地资源与环境综合调查报告》。在此基础上,蔡述明起草了《关于保护、恢复和科学利用神农架大九湖湿地资源的提案》。此提案被列为2007年省政协1号建议案。2007年"五一"期间,时任湖北省委书记俞正声,省长罗清泉带领11位省直委办厅局领导于神农架大九湖召开现场办公会,着力解决大九湖的保护和建设中存在的困难和问题。这是湖北省委、省政府为保护湿地举行的第二次现场办公会。此后,神农架大九湖成立湿地保护管理局,并获批国家级湿地公园。现在,地处鄂西大巴山地的神农架大九湖湿地公园正以它绰约的风姿迎接四海宾客。

长期生活在"优"于水而又"忧"于水的湖北水乡,对这片土地上的湿地更是倾注了蔡述明毕生的心血。看到自己的科研成果影响到政府部门决策,应用于实践并收到实效时,蔡述明倍感欣慰,感慨万分。他曾说"省委、省政府对湿地保护如此重视,先后就洪湖、大九湖的湿地保护问题合开了两次现场办公会,挽救了洪湖、推动了大九湖的生态保护建设,此举造福后代,功德无量!"

2006 年 7 月，时任湖北省政协主席王生铁（右一）和蔡述明（左一）就神农架大九湖湿地保护和科学利用开展调研

In July 2006, Wang Shengtie (first from the right), Chairman of Hubei Provincial Committee of the CPPCC and Cai Shuming (first from the left) led a research group to carry out research on wetland protection and wiseuse of Dajiu Lake in Shennongjia

2007 年时任湖北省政府副省长阮成发（左四）和蔡述明（左五）参观重点实验室

Ruan Chengfa (fourth from the left), then Vice Governor of the Hubei Provincial Government, and Cai Shuming (fifth from the left) visited the Key laboratory of CAS in 2007

为了进一步推动湿地保护与管理工作，蔡述明将其所获"拉姆萨尔湿地保护科学奖"的奖金全部捐出，并积极筹措资金建立了湖北省湿地保护基金，此举得到了湖北省委省政府的大力支持，也受到曾于2007年到访湖北的湿地公约局时任秘书长Peter Bridgewater先生的高度赞赏。

值得一提的是，蔡述明作为中国科学院的一名优秀教师，以德立身，言传身教，诲人不倦，培养了一批优秀的博士、硕士。同时，作为环境与灾害监测评估湖北省重点实验室的创始人，他所建立的湿地领域专家团队依然在湿地研究、湿地保护的道路上继续前进。

二、致力于长江湿地保护的"山水教授"——"荆楚楷模"李长安

李长安，中国地质大学（武汉）教授、博士生导师。在参加工作的40余年里，他躬身教学科研一线，特别钟情于湿地保护。

作为一位科技工作者，他长期致力于长江的形成演化、流域的生态环境保护与可持续发展研究；作为一位教师，他还积极投入长江流域生态环境保护的科普宣传；他利用人大代表和政协委员的身份，积极为湿地保护建言献策。他的足迹几乎遍布长江流域的山山水水，被称为"山水教授"，还先后被评为"武汉楷模"和"荆楚楷模"。

李长安从中国地质大学（武汉）毕业后留校任教，主要从事地貌学和第四纪地质的教学与研究工作。1998年长江流域的一场大洪水，将他的研究兴趣转向长江。当时正在西北地区科考的李长安接到了老师殷鸿福院士的电话，"老师希望我赶紧回汉，从地学角度出发，通过专业研究为长江防洪提供科学办法。"

李长安回来后，立马前往长江沿线考察，"站在大堤上，看着长江的洪水险情，我内心很震动，这是唐山大地震以后，又一次给我留下深刻印象的自然灾害。洪水的危害太大了！"于是，他开展了数年的长江流域防洪减灾研究。

他发现长江"母亲河年龄"一直存有争议，便又将研究集中于长江的形成与演化。为此，他花了近10年的时间开展了长江的考察，几乎徒步走遍了长江的各个主支流河段。

经过多年科学考察,长江流域的生态环境问题引起了李长安的关注。"长江实在太重要了,长江流域的人口数量和 GDP 占比水平都极大,其生物多样性在国际大河流域具有重要地位,是中国的重要生态环境屏障。"在大量的调查中,李长安认识到长江的伟大,也发现了长江母亲河的健康问题,又将自己研究重点转向长江的环境保护。在长江形成演化、防洪减灾、环境保护与可持续发展方面取得突出成果,曾获省部级科技成果奖 7 项,并获"全国优秀地理科技工作者"等称号。

2007 年,武汉市的湖泊环境引起了李长安的关注。"市内的大小湖泊对城市的防洪蓄水、生态环境保护和生态文明建设发挥着巨大作用。"

多年来,李长安不在上课的路上,就在调查的路上。每到双休日,李长安就带着学生开车巡湖。在每次调查前,他先是看卫星遥感影像,选择有填湖的地方及可能存在污染隐患的岸线等重点实地考察。武汉三镇的主要湖泊他基本上都做过考察。2013 年,李长安做武汉城市地质调查时,发现城市的山体环境保护也刻不容缓,很多山体因采石、挖矿遭到破坏,岩土裸露。他亲自调研了武汉市百余座大小山体,提出"加强山体保护,改善生态环境"的建议案,有效推动了武汉市的山体保护和复绿工程。

李长安教授现场讲解　Professor Li Chang'an teaching on site

江河湿地的保护，一直在李长安的心里。为写出"梁子湖生态环境保护"的建议，他驾车近十次进梁子湖调查，并呼吁设立专门的"梁子湖生态特区"，通过立法和行政手段保护梁子湖，所形成的"梁子湖报告"系列深度报道，直接推动了《梁子湖生态环境保护规划》出台。

在对长江流域的调查中他发现，长江流域经济发达、人口众多，是世界上人口密度最大的大河流域。长江的保护需要科技，更需要全民的自觉保护。于是，李长安又把大量时间投入保护长江科普宣传中。

在10余年的时间里，李长安利用业余时间深入学校、机关、社区等，开展保护长江、保护湿地、保护山水的科普讲座近400场。除了本人进行环境保护的科普宣传外，他还利用自己承担的职务——中国科协全国第四纪地质学学科首席科学传播专家、中国地质学会地貌学及第四纪地质学科学传播专家团团长、湖北省科技传播学会理事长和湖北省青少年科技协会理事长等，领导和组织科技工作者和社会资源广泛开展环境保护科普宣传工作。

李长安带领的"地貌学及第四纪地质学科学传播专家团"，曾多次获得全国先进科普团队。他在环境保护科普方面贡献突出，获得"湖北省科技传播十大杰出人物""全国科普工作先进个人"等称号，入选全国第四届"保护母亲河奖"候选人。他的长江环境保护"全民观"和"科学观"，分别写入义务教育教材《初中政治》和大学教材《环境科学概论》中。用他的话说，"科普教育能培养全民的科学精神，培养科学的思维方式，这对于民族的进步发展非常重要。"

在承担繁重的教学和科研任务的同时，李长安积极参与社会服务。他在担任中国民主促进会武汉市委员会副主委的10年期间，主要负责参政议政工作。他身先士卒，对武汉市河湖湿地的保护开展了大量的调研，形成了一系列的调研报告和提案、建议，生态环境保护已成为武汉民进参政议政工作的一大亮点。他在卸任后，武汉民进还专门成立了"李长安工作室"，《人民政协报》为此做了专版报道。

他在任全国政协委员、湖北省政协委员、武汉市人大代表及武汉市政府参事等期间，围绕长江大保护、湖北省和武汉市河、湖、湿地的生态环境保护，李长安以科学的态度，进行深入调研，积极撰写提案、建议和咨询报告达80余份。为区域环境保护、生态文明建设等发挥了重要作用，有效推动了湖北的湖泊、湿地保护，南水北调与水资源保护与合理利用以及"全国碳交易注册登记系统"落户武汉等。

他的 16 份建议获得省级以上主要领导的批示。

由于他在生态环境保护方面有突出贡献，先后获民进中央"全国参政议政成果二等奖""武汉市优秀人大代表""武汉市政府参事突出贡献奖""湖北省环保先进个人""武汉市湖泊保护先进个人"等表彰，还获得"湖北省委优秀调研成果奖"和"湖北省发展研究奖"。他的参政议政成果得到了《人民日报》《光明日报》《人民政协报》《中国环境报》《湖北日报》《长江日报》等媒体的报道。

目前，已经退休的李长安教授依然活跃在环境保护的现场。他对湖北的长江流域进行了系统调查，呼吁长江大保护要重视小流域的保护与管理，提出"山水林田湖草沙＋农户"的治理模式；建议以"农业、乡村＋生态"为特色的"梦里水乡"，打造"农旅文教与生态融合发展"的乡村振兴模式。今年，李长安教授策划和组织以"长江大保护"为主题的大学生长江源科考，67 岁的他，依然带队到海拔 5000 米以上的长江源头进行考察。

为了江河与湿地的生态环境保护，李长安教授一路砥砺前行。这些年来，他把所有的时光、知识和智慧，献给了美丽中国、美丽湖北的建设。将来，他在生态环境保护的科研与科普方面，依然"双向奔赴"，以此彰显知识分子的家国情怀与时代担当。

三、中国第一家——湖北省湿地保护基金会

湖北省湿地保护基金会于 2008 年 7 月 8 日在湖北省武汉市诞生，它是我国设立的首个湿地保护基金会。

2005 年 6 月 10 日，湿地公约局秘书处在瑞士宣布，将三年一度的国际湿地科学研究的最高奖——"拉姆萨尔湿地保护科学奖"授予中国科学家蔡述明研究员，以奖励他在湿地保护和合理利用的科学研究以及政府决策方面做出的杰出贡献，并邀请他参加在乌干达召开的《湿地公约》第九届缔约方大会。蔡述明，时任全国政协委员、湖北省政协副主席，是我国著名的湿地科学家，他在湖北省的湿地保护和环境变化研究成果也受到国内外关注。为进一步推动湖北省湿地保护与管理工作，作为国际"拉姆萨尔湿地保护科学奖"的获得者，蔡述明研究员将所获得的奖金全部捐出，并积极筹措资金建立了湖北省湿地保护基金会，此举得

湖北省湿地保护基金会成立　　Hubei Wetland Conservation Foundation is established

到了湖北省委省政府、国家林业局、民进中央的大力支持，也受到曾于 2007 年到访湖北的湿地公约局时任秘书长 Peter Bridgewater 先生的高度赞赏。

湿地保护基金会的宗旨：以习近平生态文明思想为指导，推动湖北湿地资源保护和可持续利用、促进生态文明建设为目标，鼓励和支持全社会重视湿地资源保护和可持续利用，促进人与自然和谐，为实现湖北经济社会可持续发展做出贡献。

湖北省湿地保护基金会自成立以来，在湖北省林业局、省民政厅的关心指导下，在理事单位和全体理事的大力支持下，充分发挥了湿地保护基金会的作用，积极开展湿地保护公众宣传，提供保护恢复项目的专家技术指导，保障湿地可持续利用，促进人与自然和谐，为实现湖北经济社会高质量绿色发展做出贡献。湖北省湿地保护基金会也得到了国内许多热衷于环境保护公益事业的企业支持和赞助，企业家们以实际行动为湖北湿地环保事业添砖加瓦，同时也形成了这些企业的环保文化。

湖北省湿地保护基金会现任理事长为胡兴焕；历任理事长分别为蔡述明、宋德福、侯开举。

1. 组织开展与湿地相关的普法宣讲、生态公益活动

（1）举办以《中华人民共和国湿地保护法》为主要内容的 1+N 培训活动。组

织专家采取集中和分片相结合的形式，对全省湿地保护管理人员开展《中华人民共和国湿地保护法》及湿地保护修复相关政策、项目管理等内容的培训。

（2）根据联合国第六十八次大会决议，每年的3月3日为世界"野生动植物日"。湿地基金会每年与湖北省林业局、湖北省野生动植物保护协会（以下简称湖北省野保协会）举办湖北省"世界野生动植物日"大型宣传活动。

（3）根据《中华人民共和国野生动物保护法》，每年与湖北省林业局、湖北省野保协会共同举办"爱鸟周"活动启动仪式。每年11月参与湖北省林业局、湖北省野保协会举办的"保护野生动物宣传月"等活动。并与湖北省林业局、湖北省野保协会、有关湿地公园、自然保护区共同举办了林业系统职工和志愿者的观鸟比赛活动，通过活动，加深了野生动物保护和湿地保护主管部门员工对湿地鸟类的认识，提高了业务水平，为湿地保护工作打下了坚实基础。

（4）在每年的2月2日"世界湿地日"，基金会与湖北省林业局、有关湿地公园、环保组织等单位共同举办"世界湿地日"宣传活动。在"世界湿地日"宣传活动上，对湿地保护教育工作做得好的中小学进行表彰，并授予"湖北省湿地保

为湿地保护教育学校和优秀教师授牌发证
awarding licences to wetland conservation education schools and outstanding teachers

护教育示范学校"荣誉称号。目前已授牌的学校有：武汉市华中里小学、武汉市东湖生态旅游风景区华侨城小学、武汉市大兴路小学、武汉市第二十三初级中学、洪湖市第一小学、荆门市漳河镇中心小学、荆门市沙洋县曾集镇中心小学、咸宁市咸安区官埠中学、十堰市东风22小学、南漳县东巩镇完全小学、松滋市涴水镇大岩咀小学。为了鼓励更多的中小学校教师参与湿地保护教育，湿地保护基金会对在湿地保护教育中表现突出的教师授予"湖北省湿地保护教育优秀教师"称号。

（5）由湖北省林业局、湖北省湿地保护基金会、湖北日报传媒集团、湖北省广播电视总台、世界自然基金会等单位共同发起，在全省开展首届"湖北湿地保护奖"评选活动，活动旨在弘扬生态文明，表彰和奖励那些为湖北湿地保护事业做出重大贡献和取得优异成绩的个人和集体。

（6）推进组建汉江（湖北襄阳段）湿地公园联盟。着力推进汉江中下游湿地保护修复、信息共享、提档升级、联防共治。

2. 开展湿地宣教、培训活动，提高保护意识与管理能力

（1）开展"生态湖北、大美湿地"摄影大赛活动。为了迎接2022年《湿地公约》第十四届缔约方大会在武汉举办，更好地宣传湖北省湿地保护与修复成就，提升人民群众湿地保护意识，湖北省湿地保护基金会开展了一系列活动，如：与湖北广播电视台合作，拍摄《大美湿地湖北行》电视纪录片；与新华网联合开展"生态湖北、大美湿地"摄影大赛活动。

（2）举办湿地保护教育、自然保护地管理专题培训班。湖北省湿地保护基金会先后举办了多期全省湿地保护教育师资专题培训班，一期全省国家湿地公园建设管理培训班。全省106个湿地自然保护区（湿地公园）和75所中小学的304名校长、主任、宣传干部和骨干教师报名参加了培训。培训班聘请专家、自然体验培训导师等授课。通过举办培训班，以不断提高各湿地保护区、湿地公园和湿地管理人员湿地保护知识、湿地保护技术、湿地保护管理水平。大多数学校已经将培训成果转化为教学实践。

3. 开展湿地保护修复、资源监测及专家技术支撑

近年来，通过深入调查摸底实地考察，然后邀请理事和专家组成考察组组织咨询，确定对全省七个湿地公园和自然保护区资助公益项目：

（1）神农架大九湖国家湿地公园生态监测项目。该湿地公园湿地监测具有较好的基础条件和较大的社会影响力，项目的选择在全省范围内具有唯一性、示范性和科学性。

（2）咸宁向阳湖国家湿地公园监测项目。向阳湖是我省第四大湖泊斧头湖的南半部，是江汉平原地区重要的生态节点。近年开展退田还湖、退渔还湖等系列工程，湿地水质得到明显改善，湿地生物多样性得到逐步恢复。

（3）湖北返湾湖国家湿地公园生态监测项目。返湾湖国家湿地公园属于典型的江汉平原湖泊群中的浅水型湖泊湿地，该湿地上接汉江，下连长江，是构建江汉平原乃至长江中下游湿地安全体系的重要生态区域。

（4）网湖国际重要湿地鸟类调查项目。网湖国际重要湿地地处东亚－澳大利西亚候鸟迁徙通道上，自然资源独特、地理区位优越、生物多样性丰富，鸟类种群数量在全省处于优势地位，具有十分重要的国际保护意义。加之本底调查翔实、基础工作扎实，因而开展湿地保护合作很有必要。

（5）襄阳南漳县鸳鸯及栖息地保护与监测项目。该项目位于襄阳南漳县东巩镇昌集村，2014年被中国野生动物保护协会授予"中国鸳鸯之乡"称号，是全国三大鸳鸯集中栖息地之一，在世界生物物种保护领域具有极高的生态地位和重要的物种价值。基金会支持并资助在鸳鸯湖实施鸳鸯及栖息地保护与科研监测、科普教育、资源保护、越冬食物培育等。

（6）荆门漳河国家湿地公园生产生活污水生物降解处理项目。该项目位于荆门市漳河国家湿地公园中的李集岛，项目的实施可防止岛上生活、生产污水直接排放对库区水质的污染，也为广大农村地区的生产、生活污水处理提供模式和样板。为此基金会支持漳河国家湿地公园采用人工湿地处理模式，将生活、生产污水集中到人工池塘，经过三级人工池塘的水生植物或微生物降解处理后进行无公害排放。

湖北省湿地基金会与受资助的单位共同开展湿地资源监测、候鸟监测，以参与建设并监督实施，从而打造成湖北省具有一定影响力的精品湿地生态保护示范工程。通过资助以上湿地基础研究和公益项目，湿地基金会旨在将湿地保护得好的典型宣传出去，好的技术模式作为示范案例推广。这些项目的实施取得了较好的社会经济和环境效益，同时也扩大了湖北省湿地保护基金会的社会影响，在每个项目所在地都有湖北省湿地保护基金会明显的标识牌。

4. 加强自身建设和社会参与度，提升基金会的影响力

（1）编印《湿地保护通识读本》。为了向大众普及湿地保护知识，同时为了拿出基金会自己的宣传品，基金会组织编写了《湿地保护通识读本》（以下简称《读本》）。《读本》中包括"什么是湿地""主要湿地类型""湿地的主要功能""湿地受到的威胁"和"如何保护湿地"等内容。《读本》作为基金会的宣传资料，计划在基金会参与的湿地宣传活动中向公众分发，也拟在2022年武汉举办的国际湿地大会分发，展示湖北省湿地保护基金会开展的宣传工作成效。

（2）为了充分运用互联网的广泛传播能力，基金会积极转战互联网，设立了"荆楚湿地保护"的公众号。基金会公众号分为三大板块，第一个板块介绍基金会自身；第二个板块介绍基金会开展的工作和活动；第三个板块普及湿地保护知识。现公众号构建已完成，正在进行试运行，待合适时机向公众推送。公众号还将积极寻求合作，与立志保护湿地的机构，单位及个人合作，共同经营，把"荆楚湿地保护"的公众号办成省湿地基金会的一个宣传窗口。

5. 组织社会公益募捐，管理湿地保护基金

在进一步取得社会对湿地保护重要性认知的基础上，动员更多的企业家、湿地保护的热心人士踊跃捐赠，为湖北省湿地保护事业的发展筹集更多的资金。同时，严格按照《湖北省湿地保护基金会章程》，管好、用好湿地保护基金。

四、致力于生物多样性保护的NGO——世界自然基金会

1. WWF与生物多样性保护

世界自然基金会（WWF, World Wide Fund For Nature）是在全球享有盛誉的、最大的独立性非政府环境保护组织。WWF致力于保护世界生物多样性及生物的生存环境，所有的努力都是在减少人类对生物及其生存环境的负面影响。

WWF在中国的工作始于1980年的大熊猫及其栖息地的保护，是第一个受中国政府邀请来华开展保护工作的国际非政府组织。WWF的使命是遏止地球自然环境的恶化，创造人类与自然和谐相处的美好未来。为此，我们致力保护世界生物多样性；确保可再生自然资源的可持续利用；推动降低污染和减少浪费性消费的行动。

2019 长江湿地保护网络年会
2019 Network Annual Meeting of Yangtze River Wetland Conservation

长江江豚保护大使张靓颖女士和湿地使者们
Yangtze finless porpoise conservation ambassador Ms. Zhang Lianying and wetland messengers

2. WWF 与湖北湿地保护

WWF 是第一家在中国开展自然保护的国际环保组织。从 1994 年开始，WWF 在中国开展了湿地保护与可持续利用项目，并协助国家林业局与 17 个部委制定发布了《中国湿地保护行动计划》。

世界自然基金会于 2002 年 8 月起，致力于长江中下游河湖湿地生态系统，尤其是阻隔湖泊生态系统的有效保护与湿地资源的可持续利用。在湖北，相继开展长江江豚保护、环境流、渔业市场转型、"重建长江生命之网"、"与气候伙伴同行"、"野生动植物保护小额物种"、国际湿地城市创建、国际重要湿地和国家湿地公园建设、《湿地公约》缔约方大会筹备等多项工作。

项目通过实地示范、合作网络建设及政策倡导、宣传与环境教育等手段，在推广基于社区的湿地资源保护与合理利用、重建江湖联系、长江湿地与豚类保护网络建设、流域综合管理、示范环境-气候友好型农业模式等领域开展了富有成效工作，为各级政府和当地社区提供了诸多行之有效的解决办法和实际案例。

2004 年，湖北省林业局和 WWF 合作共建"湖北省湿地保护区网络"，探索运用湿地保护区网络建设的手段来提高湿地有效管理的方法和途径。在此合作模式的基础上，在国家林业局（现为国家林业和草原局）和长江流域各省市政府共同支持下，创建流域层次的网络平台——长江湿地网络，该网络旨在推动湿地的保护与可持续利用，共同应对全球气候变化对区域可持续发展带来的影响，提高了流域内湿地保护管理的能力。

五、湿地保护其他社会组织和人物

1. 湖北省野生动植物保护协会

湖北省野生动植物保护协会在野生动植物科普、未成年人生态道德教育、生态文化传播等方面做了大量工作，产生了良好的社会影响，为加强生物多样性保护，推进生态文明建设做出积极贡献。

协会主要围绕加强野生动植物保护宣传、开展未成年人生态道德教育、传播生态文化、协助政府部门开展野保工作等方面开展工作。协会先后主办了 4 次世界野生动植物日、4 次全省爱鸟周、4 次全省保护野生动物宣传活动；出版发行了

湖北省野生动植物保护协会第五次会员大会
The Fifth General Meeting of Hubei Wildlife Conservation Association

4部未成年人生态道德教育书籍，编印了1部未成年人生态道德教育经验汇编；开展了形式多样的护飞活动，救助鸟类80余种，1200余只；举办候鸟护飞行动进校园、进社区、进乡村、进家庭科普宣讲活动100余场，发放宣传资料逾10000份。协会各项工作扎实推进，并获得多项荣誉。3位会员获得国家林业局（现为国家林业和草原局）保护森林和野生动物资源先进个人，8位会员被中国野生动物保护协会授予"斯巴鲁生态保护奖"。

2. 武汉观鸟会——志愿者之家

武汉市观鸟协会是致力于野生鸟类研究、保护和生态环境保护的社会公益团体。2017年3月30日在武汉市民政局正式注册，业务主管单位为武汉市园林和林业局。

多年来，协会一直坚持以志愿者服务的方式，在武汉开展自然教育、鸟类监测、鸟类研究、鸟类救助和救护，以及鸟类栖息地保护。"在一起，把有意思变得有意义"凝聚了全体会员的共识，促进人与自然和谐共生是协会为之奋斗的目标。

2016年7月，协会动员以会员为主的志愿者，在武汉地区开展重点区域鸟类监测公益项目，聚集观鸟爱好者的力量进行有规律的鸟类观测、科学研究和鸟类及

鸟类监测
bird monitoring

保护宣传
conservation advocacy

鸟类救助
bird rescue

栖息地保护。研究武汉地区鸟类分类与分布、监测重点区域地块鸟类动态变化、发现威胁鸟类生存因素，是重点区域鸟类监测公益项目的主要目标。至2021年底，在志愿者的努力下，设立的70个重点区域鸟类监测点遍布武汉三镇，其中包括蔡甸沉湖国际重要湿地，新洲涨渡湖、江夏上涉湖、黄陂草湖、汉南武湖等省、市级湿地自然保护区，武汉东湖、蔡甸后官湖、江夏藏龙岛、江夏安山、东西湖金银湖、东西湖杜公湖等国家湿地公园以及武汉马鞍山、江夏八分山、黄陂木兰山等城市森林公园，涵盖了湿地、森林、农田、城市绿地、居民区等各类生境类型。

协会在5年中共收到监测报告9288份，向社会公布鸟类监测月报64期，连续6年编制鸟类监测年报。通过采集、统计、总结、发布武汉市的鸟类的分类、分布及动态变化，为武汉市生态环境保护积累了大量的科学数据，为武汉市的生态文明建设乃至长江生态大保护做出了积极的贡献。

在协会的动员和鼓励下，部分会员持续开展鸟类救助行动，5年中共实施鸟类救助2348次，救助野生鸟类约17目41科134种，共3564只。其中包括国家二级重点保护鸟类共23种，省级重点保护鸟类共30种。

协会一直坚持开展自然教育公益服务和鸟类保护知识宣传，在东湖绿道、湿地公园、森林公园、社区等地开展以湿地和候鸟保护为主题的免费自然教育活动128场；走进中小学，举办线下讲座，活动覆盖约4万人次。

协会编撰出版了《武汉鸟类图鉴》，该书总结武汉市20年鸟类观测成果，全面介绍截至2020年有记载并确认的419种野生鸟类，首次正式发布武汉鸟类名录，为武汉市鸟类资源交出重要的科学研究成果，为市民鸟类观察、自然教育提供普及翔实的学习工具。《武汉鸟类图鉴》获得2021年度武汉市"优秀科普图书奖"。协会还编撰出版了《窗外的鸟：武汉宅家观鸟报告》，该书是在2020年初武汉抗疫特殊时期人与自然深度交流的美好产物，也是人与自然和谐共生的力量传递。《窗外的鸟：武汉宅家观鸟报告》荣获第二届"坪山自然博物图书奖读者推荐大奖"。

3.湿地学校——生态教育进校园

（1）爱鸟护湖卅三载，湿地教育树标杆——武汉华中里小学

华中里小学秉承"适性共生，适切竞长"的办学理念，从1989年就开始了湿地教育之旅。

起步：护鸟为媒，特色育人。学校"爱鸟护鸟"教育始于1989年，并成立了

华中里小学爱鸟小分队
bird care team of Huazhongli Primary School

校园湿地
wetlands in school

走进湖泊经典诵读
classic recitation to be close to lakes

全国最早的学生爱鸟社团。学校将爱鸟护鸟教育与学科课程、德育活动、传统文化等内容相结合，并开发《鸟语唐诗》校本教材。每逢"爱鸟周"，学校都会开展丰富多彩的爱鸟护鸟活动，通过讲鸟、舞鸟、唱鸟、画鸟、写鸟、观鸟、制作鸟巢、倡议护鸟等多种形式表达对鸟的热爱之情，向市民宣传保护鸟类的重要性，师生的观鸟护鸟足迹不仅遍布武汉市区湿地公园，还往外拓展到武汉周边的沉湖和京山。

基地：校园湿地，常态宣教。多年的"爱鸟护鸟"行动让华小人明白一个道理：单纯地爱鸟护鸟是不够的；要想保护好鸟，就要保护好鸟儿的栖息地——湿地。2006年，学校按照"总体规划，逐步投入，逐步建设"的思路，先后投入100余万元建立了全国第一家小学湿地生态教育馆和屋顶湿地花园，开发了《湿地——生命的摇篮》校本课程，让学校的湿地教育进入常态宣教阶段。2009年，该校被湿地国际中国办事处授予"湿地实验学校"称号；因湿地教育特色，学校承办了第十三届世界湖泊大会中日韩青少年分会场活动，华中里小学再次因爱鸟护湖的特色成为人们关注的焦点。

拓展：校外调查，深入实践。2015年，为进一步丰富湿地教育的内涵，学校结合武汉"百湖之市"的特点，从武汉市166个湖泊中选取具有典型代表的14个湖泊开发了《美丽武汉 多姿湖泊》校本教材。教材从湖之风姿、湖之历史、湖之名胜、湖之今朝、湖之诗词歌赋、游湖感言等方面引导学生关注武汉本地的湖泊湿地，感受家乡湖泊的变化，激发爱湖之情，生发护湖之行动。2017年起，学校利用暑假以"关注湿地 爱我百湖 争做护湖小使者"为主题先后策划了的"爱我百湖——都市绿肺""游湖赏荷""我家门口的湖泊"等系列调查实践活动，组织学生走进武汉市的各大湖泊进行湿地考察、采访市民、撰写科普手记……主题活动从校内到校外，再从校外到校内，从知识学习到实地体验，再到保护行动，多层次、多角度地推进湿地教育。2018年，学校被湖北省湿地基金会评为湖北省湿地保护示范学校，被生态环境部宣传教育中心批为全国首批"美丽中国，我是行动者——小河长小湖长"青少年环境志愿者行动试点学校。

坚守：诗意前行，护湖标杆。从成立全国第一个小学生爱鸟护鸟社团，到建立全国第一个小学湿地生态教育馆，到先后成为全国第一批国际湿地学校、全国首批"美丽中国，我是行动者——小河长小湖长"青少年环境志愿者行动试点学校、全国第一批国际生态学校，"爱鸟护湖"，学校坚守了33年。从护鸟到湿地教育，从校内到校外，从常态宣教到深入实践，"绿"成了学校浓墨重彩的底色，"让鸟儿有个家"成了全校师生的庄重宣言。春有观鸟，夏有研湖，秋有自然体验，冬有候鸟监测……学校以"爱鸟护湖"的小行动，成就着生态文明的大奇迹，逐渐成为华中地区唯一坚守33年的爱鸟护湖标杆学校。

（2）拥抱绿色东湖，缤纷多彩童年——华中师范大学附属华侨城小学

华中师范大学附属华侨城小学坐落于5A级国家生态旅游风景区东湖北岸，学校依托得天独厚的东湖生态环境，重视生态文明建设，注重培养师生环境意识，让生态文明的种子深入孩子们的心灵，内化于心，外化于行。

学校以"生态东湖"为切入点，确立"生态文明教育"办学特色，构建"华侨城小学教育生态系统"，将东湖生态环境汇聚成课，通过问答、游戏、图画以及实践活动深入浅出地引导孩子探究式学习，认识东湖生态系统，形成一种新的生态自然观生态道德观，并倡导身体力行保护生态环境。

生态环境保护教育　ecological and environment conservation education

湿地水质监测　wetland water quality monitoring

环保小卫士　little environment guardians

研发教材，渗透湿地生态文明理念。华侨城小学依照新课程标准的要求，把湿地生态文明教育渗透进语文、数学、科学、美术等学科教学的全过程，充分挖掘教材中的低碳环保教育因素，学科融合，树立生态环境教育意识。学校共编写了三本校本教材《魅力东湖》《人文东湖》《生态东湖》；通过承担全国少先队和湖北省教育科学规划的研究课题，深化湿地保护教育内涵。

建设基地，增强湿地生态文明意识。一是校内东湖生态文明教育馆，把"生态文明教育"这个大题目、大文章，从"小"处入题，聚焦"生态东湖"，分别展示东湖城市绿心、湿地保护、水质优化与环境治理、生物多样化等方面的成就，激发学生了解东湖、热爱东湖、保护东湖、建设东湖，通过声光电技术，为师生提供大视野近距离与东湖交流的新天地；二是校外华阳湖社会实践活动基地，以小见大，师生共同在华阳湖的小湿地环境中学习具体科学保护大东湖湿地的方法。

实践活动，共创湿地生态文明家园。一是定期巡湖，学生、家长、老师们随时关注东湖的环境变化，拍照摄影捕捉东湖变化，用笔记记录东湖自然美景，身体力行保护东湖湿地；二是东湖水质的监测，每两周调查分析导致水质变化的原因，关注东湖湿地水资源环境现状；三是建立小湖长制。学生开发《华阳湖》校本教材、学生组建生态宣讲团、阳光家长环保讲坛、专家进校园、走访科学家、"拥抱绿色 缤纷童年"生态艺术节、"绿 LOVE"低碳环保运动会、世界环境日宣传活动等。

（3）心系汉江水，浓浓湿地情——武汉市大兴路小学

曾获得联合国环境规划署"全球 500 佳"青少年环境奖的武汉市大兴路小学，20 多年前就因环境教育而"绿"扬天下，但学校的"绿色"长征从未止步，一直在寻求创新与突破。从 2012 年开始，大兴路小学涉足湿地教育，持续开展汉江生态考察活动。曾经用五年时间，带领学生考察汉江，极富特色的生态实践活动在孩子们的成长过程中写下浓重一笔，也开辟了生态教育的崭新路径。

学校环境教育与时俱进。由于毗邻长江汉江交汇处，与汉口龙王庙咫尺之遥的地域优势，学校将视野投向荆楚大地的"生命之源"——汉江。2012 年 6 月，大兴路小学在汉口龙王庙成立了汉江流域首个水环境教育校外实践基地，开启河流湿地研究。

2015 年 8 月，学校以"汉江情　朱鹮缘"为主题，对接中日合作人与朱鹮和

谐共存的地区环境建设项目，依托汉江考察开展生态道德教育环境实践活动。

大兴路小学的汉江考察成果得到了广泛的关注和认可。2014年，学校获得湖北省"环境保护政府奖"，2016年，学校荣获团中央等中央八部委授予的第七届"母亲河奖"，这些都为学校深化汉江生态考察活动提供了无限动力。

学校的考察汉口活动将每五年一次，第一阶段的考察计划结束了，第二阶段的考察活动将重点聚焦于南水北调和汉江经济带发展大背景下，汉江水质的变化与河流湿地保护等。如今，这项考察活动已成为大兴路小学的特色品牌，为湿地保护实践树立起一个醒目的标杆。

但是科学考察的旅程远没有到达尽头，推动湿地教育的持续发展依旧是一个长远的课题。对此，大兴路小学的探索将永不止步。

（4）印记长江濒危鱼种——武汉市第十二中学

在武汉市第十二中学陈列室，一幅幅黑白分明的版画，清晰地记录着长江濒危鱼类的"身份信息"。这些是自2018年起，该校柳娜老师带领同学们开展的"印记长江濒危鱼种"项目中的作品，一条条濒危的长江鱼种在同学们刻刀下重新焕发出生机和活力。

刻印长江濒危鱼种版画，源于学校开展的一场讲座——《留住江豚的微笑》。2018年暑期，来自青岛的八所学校和武汉市第十二中学"光影社"的同学循着这场讲座的足迹开始了一场"寻豚之旅"。他们一道参观中科院水生生物博物馆，向白鱀豚自然保护基金会捐款，并对采集到的20袋江豚栖息地泥土样进行检测，了解酸碱性、成分、是否含重金属以及分析江豚的生活环境等。

武汉市第十二中学"光影"根与芽小组发起了《守护长江精灵 印记濒危鱼种》项目。同学们将水中"沉默的消亡者"印成版画展现在人们面前,目前已经完成了超过一百多种长江鱼种的版画创作,我们的目标是建立长江鱼种的"全家福"。

自《守护长江精灵 印记濒危鱼种》项目开展以来，他们经常带着精心制作的版画，在不同的地方邀请人们一起来印制。在武汉的长江文明馆、武汉科技馆、汉口菱角湖广场、解放公园的科普活动现场，都出现过同学们活跃的身影；2021年，师生们还受邀参加在云南昆明举办的COP15非政府组织平行论坛，为维护长江鱼种多样性发声。

汉江生态考察
ecological investigation of Hanjiang River

湿地鸟类观测
wetland bird monitoring

长江濒危鱼种版画展示
imprinting display of endangered fish species in the Yangtze River

4. 湿地保护一线的管理人员、志愿者

（1）冯江——湖北蔡甸沉湖湿地自然保护区

冯江大学毕业以后，积极报名参加"三支一扶"计划，来到武汉市蔡甸区沉湖湿地自然保护区管理局工作，从此在生态环保战场上奋斗已10多年；他风雨无阻，坚持每天骑车、步行进行巡护监测。他创造了保护区多个记录：驻守保护区基层站点时间最长的人，保护区认识鸟类和植物最多的人，保护区发现鸟类新纪录最多的人，保护区的"活地图"，跑遍了保护区内外所有乡镇、村组。

他获得了多项荣誉：2010年武汉市"爱鸟护鸟先进个人"，2013年"湖北省自然保护区建设管理工作优秀个人"，2016年武汉市"文明市民"，2017年"湖北省最美野生动物守护人"提名奖、2019年"最美基层高校毕业生"提名奖，2019年"武汉楷模"，2021年湖北省"爱鸟护鸟先进个人"。

（2）杨涛——湖北石首麋鹿国家级自然保护区

杨涛于2011年进入石首麋鹿国家级自然保护区工作，主要从事保护区的科研监测工作。作为一名从专业院校毕业的大学生，他耐得住寂寞，扎根保护区。他勤奋学习，努力做好巡护监测工作。制定监测方案，建立了保护区生物多样性监测体系，摸清保护区内野生动植物"家底"。他热爱事业，全心全意救护脱群小麋鹿。在专家和同事的帮助下成功救护受伤小鹿，并已放归野生麋鹿种群。他科学建言，加强湿地生境修复和保护。积极参与综合科学考察和成果总结，还独立或合作撰写发表了《湖北石首天鹅洲故道湿地鸟类多样性研究》《石首麋鹿保护区麋鹿疫病防控现状及措施》等报告，为保护区的保护恢复与管理提供建议。

（3）丁泽良——湖北长江天鹅洲白鱀豚国家级自然保护区

丁泽良目前主要负责天鹅洲故道网箱饲养长江江豚的训练、繁殖和管理工作，同时兼顾天鹅洲故道周边环境、渔业资源以及江豚种群的监测、巡护、考察等工作。丁泽良原为天鹅洲故道专业渔民，自幼生活在长江边，自2008年开始正式受聘于长江天鹅洲白鱀豚国家级自然保护区，并作为最重要的渔民技术力量，参与了历次中国科学院水生生物研究所组织的长江江豚种群调查以及长江江豚迁地保护工作。

（4）颜军——武汉观鸟协会

颜军，鸟类保育工作者，武汉市自然教育专家。武汉市观鸟协会（武汉观鸟会）创始人之一，现任湖北省野生动植物保护协会常务理事，武汉市观鸟协会会

261

第五篇
湖北湿地保护典型人物与组织

冯江　Feng Jiang

杨涛　Yang Tao

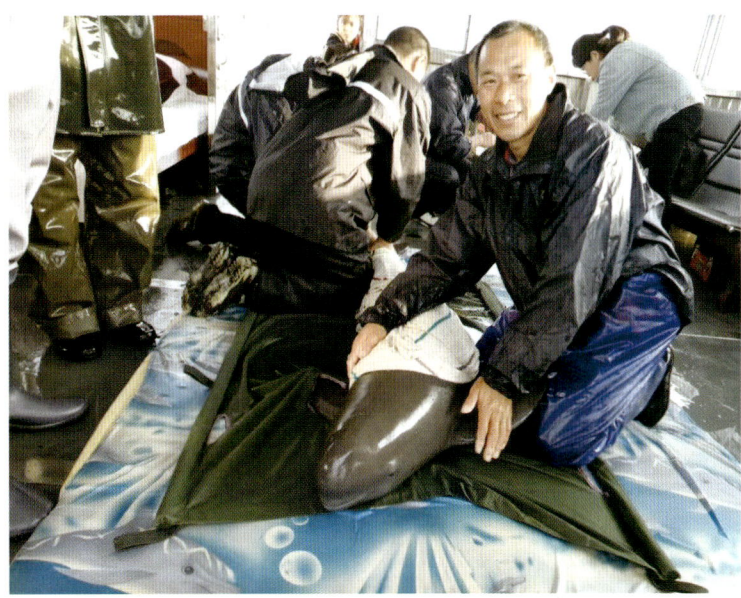

丁泽良　Ding Zeliang

颜军　Yan Jun

汪德新（中）　Wang Dexin (middle)

长，武汉市公园协会自然教育分会常务理事。

15年来，他一直坚持在武汉及周边地区观察、拍摄、研究野生鸟类，推广观鸟活动，在政府有关部门指导下开展鸟类调查监测、鸟类保护和栖息地保护。曾发表《武汉市黄陂区青头潜鸭的分布初步研究》等十多篇鸟类研究论文；连续六年编撰出版《武汉重点区域鸟类监测年报》；主编《武汉鸟类图鉴》和《观鸟手册》；主编的《窗外的鸟：武汉宅家观鸟报告》荣获第七届"中国科普作家协会优秀科普作品奖"科普图书金奖。2017年被评为湖北省"爱鸟护鸟先进个人"，2018年被评为全国志愿者"志愿者护飞行动先进个人"，2019年被评为武汉市"最美志愿者"。

（5）汪德新——湖北远安沮河国家湿地公园义务巡护员

汪德新是湿地公园的守护者中的一员，湖北远安沮河国家湿地公园义务巡护员，默默守护着这个绿色家园。汪德新家就在沮河国家湿地公园的旁边，是2018—2021年湿地公园聘用的公益性岗位巡护员之一。2022年湿地公园取消公益性岗位后，汪德新就自愿成了一名义务巡护员，经常为周边居民和市民游客科普宣教。

六、湖北湿地保护的科研支撑

灵秀的湖北人杰地灵，更是拥有雄厚的科研教育资源，是名副其实的科教大省，湖北省的高校在质量和数量上均名列全国前茅，同时还拥有数量众多的科研院所。依托于科研院所和高校良好的科研平台，构建了多个湿地科学相关的实验室和研究中心，并为湖北省湿地保护与监测相关科研工作的开展提供良好的平台。

湖北省积极参与国际国内湿地保护合作与交流，开展对湿地的科学监测与研究，持续不断提高科研水平。中国科学院精密测量科学与技术创新研究院、中国科学院水生生物研究所、武汉大学、中国科学院武汉植物园、中国地质大学等高等院校、科研单位对洪湖、梁子湖、天鹅洲故道、沉湖、龙感湖、神农架大九湖等湖北重要湿地进行生态定位观测研究；在湖泊、沼泽湿地的研究，湿地污染的防治，长江江豚、中华鲟等珍稀水生动物的保护、研究及科技合作、支撑与交流等方面做了大量的工作，并且取得了丰硕的科研成果。

1. 中国科学院精密测量科学与技术创新研究院

中国科学院精密测量科学与技术创新研究院及湖北省环境与灾害监测评估重点实验室，自成立以来，承担了国家和地方政府的科研项目，在长江流域湿地保护和水生态治理方面取得了一系列理论与应用科研成果。出版了湖北省学术著作专项资金项目——《湖北湿地生态保护研究丛书》（10册），是湖北省首套湿地保护研究系列丛书。研究院科学家蔡述明荣获2005年《湿地公约》拉姆萨尔湿地保护科学奖。实验室将遥感技术与定位实验方法应用于湿地生态系统的演化过程与动态监测，基于湿地系统生态过程与环境效应，研究受损湿地的退化机制与恢复、重建理论与技术等。研究院拥有中国科学院洪湖湿地生态系统野外科学观测研究站，

中科院野外科学观测研究站　The Field Scientific Observatory and Research Station, CAS

中科院东湖湖泊生态系统试验站
The Experiment Station of Lakes Ecosystem of East Lake, CAS

湖北省江汉平原湿地生态系统野外科学观测研究站，湖北省面源污染防治工程技术研究中心。多年来持续在洪湖、长湖、大九湖、梁子湖、网湖、龙感湖、沉湖、汤逊湖、保安湖、长江干支流开展了大量湿地保护相关科研工作，持续建言献策，为湖北省湿地保护工作的开展及保护体系的完善提供了不可或缺的科学支撑。

2. 中国科学院水生生物研究所

中国科学院水生生物研究所是从事内陆水体生命过程、生态环境保护与生物资源利用研究的综合性学术研究机构。水生所面向国家重大战略需求，围绕内陆水体生命过程、生态环境保护与生物资源利用领域的基础性、战略性和前瞻性重大科技问题，在水环境保护、淡水渔业和微藻生物技术领域发挥引领示范作用。水生所拥有湖北东湖湖泊生态系统国家野外科学观测研究站、三峡水库香溪河生态系统实验站等共建机构及分析测试中心平台。中国科学院院士曹文宣是"长江十年禁渔"首倡者，"长江十年禁渔"已经成为"长江大保护"国家战略中的重要行动计划。水生所自成立以来共有200项成果获得奖励，一些重要技术创新或技术集成获得成功并得到应用，对于湖北省及全国水环境保护和渔业可持续发展起到了显著的推动作用。

3. 武汉大学梁子湖湖泊生态系统国家野外科学观测研究站

武汉大学梁子湖国家野外生态站以长江中下游草型湖泊和水生植被为研究对象，探索自然灾害和环境污染对湖泊生态系统的影响机理，为我国湖泊生态环境保护和受损淡水生态系统恢复提供第一手数据资料和技术支撑。武汉大学1992年开始在梁子湖进行水生生态系统长期定位观测和退化水生植被重建试验，成功种植水生植被20万亩，为湖泊水质改善提供了基础保障和技术支持。

湖北梁子湖湖泊生态系统国家野外科学观测研究站，是全国第一个以水生植物和湖泊生态环境保护为研究对象的国家站，服务于国家水污染治理和湖北省湖泊保护的需求，力争为梁子湖生态环境保护与恢复做出应有贡献，为全国日益严重的湖泊富营养化治理与生态修复提供示范模式。

4. 中国科学院武汉植物园和中国地质大学（武汉）湿地演化与生态恢复湖北省重点实验室

湿地演化与生态恢复湖北省重点实验室成立于2008年10月，是由中科院武汉植物园和中国地质大学（武汉）联合组建的。实验室瞄准学科前沿，强调地理学、地质学、生物学与生态学等领域的学科交叉，其研究工作是以湖北省湿地生态系统的可持续发展为总体目标，结合湖北省湿地生态系统的结构与功能特点、受威胁情况，以及生物多样性保护的特点，开展相关监测和研究工作，包括湿地生态系统的动态监测、生物学和生态学基础，以及受损湿地生态系统的恢复与重建的理论与方法等。

5. 湖北省林业科学研究院（中国林业科学研究院湖北分院）

湖北省林业科学研究院成立于1959年，是以应用研究和技术开发为主的省级林业综合性公益型科研单位。2011年挂牌成立中国林业科学研究院湖北分院。林科院始终坚持围绕全省的生态经济发展战略和"生态湖北，绿色崛起"的发展理念，积极开展科技攻关、推广和服务。

在湿地保护修复领域，研究院2014年获国家林业局批准成立湖北洪湖湿地生态系统定位研究站，2021年经省林业局批准正式成立湿地研究所，2022年经国家林业和草原局批准牵头筹建"长江中游湿地保护修复国家创新联盟"。开展全省湿地生物多样性、湿地生态系统服务、湿地生态恢复与重建，湿地保护与管理技术研究、湿地生态及湿地碳汇/源监测等方面的科学研究与技术创新工作。

6. 湖北省野生动植物保护总站

湖北省野生动植物保护总站负责全省野生动植物、自然保护区和湿地资源调查、监测以及自然保护区网络建设；组织实施了全球环境基金资助的"生命主流——加强湿地保护体系，保护生物多样性"子项目"加强湖北省湿地保护体系，保护生物多样性项目"，通过完善省级政策框架、优化流域级管理体系、加强自然保护区能力建设，提高了湖北省湿地保护体系管理的有效性，保护具有全球重要意义的生物多样性和湿地生态系统。并及时总结各类项目成果，出版了《湖北湿地》《湖北省水鸟资源与保护》《湖北省湿地自然保护区监测手册》《沉湖国际重要湿地生物多样性及生态价值评估研究》《湖北省湿地生物多样性保护及自然保护区建设》《跟着湿地去旅行》等书，《跟着湿地去旅行》于2019年荣获湖北省"优秀科普作品"。

除此之外，长江水利委员会长江科学院、长江水利委员会长江水资源保护科学研究所、华中科技大学、华中农业大学、湖北大学和江汉大学等科研机构和高等院校在湖北湿地资源监测、湿地生态修复、湿地保护体系研究、湿地生物多样性保护等方面也开展了大量的基础研究工作，并取得了丰富成效，为湖北省湿地保护积蓄了持久弥新的科研支撑力量。

Chapter 5

Typical Figures and Organizations of Wetland Conservation in Hubei

Hubei has been exploring and pushing forward wetland conservation and construction along the course of following the harmonious development of human and nature and society. Generations of conservationists work hard to innovate and pioneer with silent dedication, resulting in fruitful achievements of wetland conservation in Hubei, and a beautiful picture of the harmonious coexistence of human and nature that is slowly unfolding itself.

In this course, millions of personnel who have been silently dedicating themselves at the first line of the ecological conservation and management have become a beautiful scenery of that beautiful landscape of Hubei. Among them, there are famous scientists and professors engaging in wetland conservation who have been working diligently for decades and appealing for support to promote decisions of national and local governments; there are wetland managers in the front line of conservation and management who have been dedicating the most beautiful years of their youth to the mountains, forests, grasslands, and wetlands in Hubei; there are guardians and volunteers of nature reserves who have been taking ecological conservation of wetlands as their highest pursuit; meanwhile there are some social environmental conservation organizations who have been actively involving in the wetland conservation in Hubei to spread the concept of harmonious coexistence to thousands of families and benefit thousands of households.

5.1 Cai Shuming, a famous wetland scientist exerting lifelong effort to guard the "kidneys of the earth"

"The study of the evolution of lakes in the middle reaches of the Yangtze River and the conservation and utilization of wetlands have been my lifelong profession. I studied physical geography in the Department of Geology and Geography in college, worked at the Institute of Geodesy and Geophysics of the Chinese Academy of Sciences, and I came to work in Hubei in the 1960s. Hubei is the 'province of a thousand lakes', and Wuhan is the 'city of a hundred lakes'. It is not far from the Dongting Lake and Poyang Lake, and there are unique conditions to study wetlands here. Wetlands are the wealth given by nature to human beings. They are so beautiful that it is pleasant to work in the field. It is lucky to deal with wetlands all my life."

-- Cai Shuming

Chapter 5
Typical Figures and Organizations of Wetland Conservation in Hubei

On November 8, 2005, at the 9th Conference of the Contracting Parties to the Ramsar Convention in Uganda, a scientist from China received the Ramsar Wetland Scientific Award, the highest award of the International Ramsar Convention. This is Cai Shuming, the famous Chinese wetland scientist. The Ramsar Award was established in 1996, and Cai Shuming is the first scientist from China to receive this honor since the establishment.

On March 5, 2002, Zhu Rongji, the then Premier solemnly proposed "strengthening wetland conservation" in the *Report on the Work of the Government*. A senior expert's mood was unsettled off the stage. His name is Cai Shuming, a member of the National Committee of the Chinese People's Political Consultative Conference (CPPCC), then vice chairman of Hubei Provincial Committee of the CPPCC, a researcher at the Chinese Academy of Sciences. He said excitedly, "This is the first time. None of the previous *Report on the Work of the Government* specifically mentioned 'strengthening wetland conservation', indicating that the country is paying more and more attention to wetland issues." When the two sessions were held in 2000, Cai Shuming had called on the state to pay attention to the conservation of wetlands, and in the two sessions in 2002 he again proposed that the state should formulate a "wetland conservation law" as soon as possible to ensure the sustainable development of the national economy and society.

Since the 1960s when he graduated from university and set foot on the Jianghan Plain, Cai Shuming has been devoting to wetlands: studying them, understanding them, and protecting them. In more than fifty years of dealing with wetlands, Cai Shuming has always been full of enthusiasm and does not find wetland research a dull affair at all. "Wearing a straw hat, hanging binoculars, sitting on the wobbly expedition boat, marching in the lakes, reeds, marshes, I conduct wetland fieldwork all year round regardless of the heat and cold. Once back to the laboratory of research unit, I analyze and collate as soon as possible the first-hand data to find out the law governing the environmental changes of wetlands." This is the daily work of Cai Shuming.

Through his long-term efforts and dedicated research in the field of wetland science, he has obtained a series of achievements: (1) systematically studying the environmental evolution of the Jianghan lake groups and Dongting Lake in the middle reaches of the Yangtze River; (2) innovatively proposing the theory of positive and negative effects of lake swamping in north subtropical regions, the findings of which and effective measures of prevention and control have been adopted by national policy-making departments; (3) proposing an agricultural development model suitable for the ecological characteristics of the wetlands in the Yangtze River basin, and its results have been applied.

In 1997, Cai Shuming was appointed Vice Chairman of the Hubei Provincial Committee of the Chinese People's Political Consultative Conference and a member of the National Committee of the Chinese People's Political Consultative Conference. From then on, he shouldered the mission of a CPPCC member, and his career also embarked on a new journey of scientific advocacy and wetland conservation.

In 1998, as a member of CPPCC National Committee and a scientist, he made several field visits to the plain lake areas of Hubei, Hunan and Jiangxi provinces in the middle reaches of the Yangtze River to support and promote the conservation of wetlands in the middle reaches of the Yangtze River. He worked hard to promote the establishment and construction of the national nature reserve of Honghu Lake in the Jianghan Plain while also put forward positive opinions and suggestions on the construction of national nature reserves in the Dongting Lake and Poyang Lake.

As the 9th and 10th CPPCC member, he has made many comments and suggestions at the meetings of the CPPCC, calling on the state to protect wetlands and pay attention to the construction of ecological environment of wetlands.

In 2002, at the fifth session of the ninth CPPCC National Committee, he put forward a proposal on the conservation and rational use of wetlands in the middle reaches of the Yangtze River, calling for the conservation and rational use of wetlands to be incorporated into the legal system. He also made a speech on "state should formulate a wetland conservation law as soon as possible to ensure the sustainable development of the national economy and society". His proposal on wetland conservation became an important basis for the wetland legislation of the National People's Congress later.

In the same year, as a member of the China Association for Promoting Democracy, he drafted a proposal on wetland conservation and wetland legislation for the Central Committee of the Association. This proposal was adopted by the then president of the country. Wetland conservation was also added to the Premier's Report on the Work of the Government that year, which caused a great response at home and abroad and played a decisive role in wetland conservation.

In 2004, at the second session of the 10th CPPCC National Committee, he proposed to build a national nature reserve in the Danjiangkou reservoir area in the upper reaches of the Hanjiang River, which is the water source of the medium route project of the national South-to-North Water Diversion in order to ensure that "a reservoir of clean water is sent to Beijing". In May of the same year, as a member of CPPCC National Committee, he accompanied the Standing Committee of the CPPCC to conduct a special study on water security for the South-to-North Water Diversion Project and visited Shiyan and Danjiangkou to investigate soil loss. At that time, he made six suggestions which were adopted in the *Research Report on Water Security of the Medium Route Project of South-to-North Water Diversion* submitted by the CPPCC National Committee to the CPC Central Committee and the State Council.

In October 2004, he participated in and chaired the Fifth Seminar on Water Environment of the Yangtze River by Provincial and Municipal Committees of CPPCC of the Yangtze River Basin. After the meeting, he submitted a proposal on water environment conservation for 13 Provincial and Municipal committees of CPPCC of the Yangtze River basin to the State Council in a timely manner, which was highly evaluated by the CPPCC.

On the basis of several in-depth investigations and studies of Honghu Lake, he summarized and prepared the *Report on Comprehensive Scientific Investigation of Wetland Nature Reserve in Honghu Hubei*. Cai Shuming was very worried about the damage to the ecological environment of the wetlands of Honghu Lake, so he submitted a proposal on *Optimizing the Management of Honghu Lake and Conserving Wetland Resources* to comrade Yu Zhengsheng, then member of the political bureau of the CPC central committee and secretary of Hubei Provincial Party Committee, which attracted the high attention of the Provincial Party Committee and Provincial Government. On November 29, 2004, the Hubei Provincial Party Committee and the Provincial Government held a meeting in Honghu of "strengthening the ecological conservation of Honghu Lake". Yu Zhengsheng, the then secretary of Hubei provincial party committee, the governor Luo Qingquan, other leaders of the provincial party committee and provincial government, leaders of more than 20 provincial offices and departments and leaders of Jingzhou city, Honghu city and Jianli county attended the meeting. At the meeting, secretary Yu made an important speech and put forward the guiding ideology and specific measures to treat Honghu Lake. This is the first on-site office meeting of such a high level on the ecological conservation of wetlands in Hubei Province, and even uncommon in the country, which has caused a strong social repercussions.

In July 2006, Cai Shuming and his team conducted a comprehensive scientific survey of the Dajiu Lake in Shennongjia and wrote the Report on the Comprehensive Survey of Wetland Resources and Environment of the Dajiu Lake in Shennongjia. On top of that, Cai Shuming drafted a *Proposal on Conservation, Restoration and Scientific Use of Wetland Resources in the Dajiu Lake in Shennongjia*. This proposal was listed as the No. 1 proposal of the provincial CPPCC in 2007. During May Day holiday 2007, Yu Zhengsheng, the then secretary of Hubei Provincial Party Committee and governor Luo Qingquan led the leaders of 11 departments and bureaus to hold an on-site office meeting at the Dajiu Lake in Shennongjia to solve the difficulties and problems in the conservation and construction of the lake. This is the second on-site office meeting held by the Hubei provincial party committee and the provincial government for the conservation of wetlands. After that, a wetland conservation administration was established for the Dajiu Lake in Shennongjia, and was approved as a national wetland park. Now, the Shennongjia Dajiu Lake Wetland Park, located in the Daba Mountains in the western Hubei, is welcoming guests from all over the world with her graceful posture.

Cai Shuming lives a long life about the "superiority" in the water and "worry" in the water in the water village in Hubei, and he also makes a lifelong commitment to the wetlands on this land. He feels gratified and deeply moved when he sees the results of their own scientific research influence the decision of government departments, applied to practice and receive effectiveness. He once said, "The Provincial Party Committee and the Provincial Government pay so much attention to the conservation of wetlands that they have held two on-site office meetings on wetland conservation issues on Honghu Lake and Dajiuhu Lake to save Honghu Lake and promote the construction of ecological conservation of the Dajiuhu Lake. This act will benefit future generations and will be of immense merit!"

In order to further promote wetland conservation and management, researcher Cai Shuming donated all the award money of the Ramsar Wetland Scientific Award and actively raised funds to establish the Hubei Provincial Wetland Conservation Fund, which was strongly supported by the Provincial Government of Hubei Province, and was also highly praised by Mr. Peter Bridgewater, secretary general of the Ramsar Convention Bureau, who visited Hubei in 2007.

It is worth mentioning that Cai Shuming, as an outstanding teacher with Chinese Academy of Sciences, has established himself with virtue, taught by example and with tireless zeal, training a number of excellent doctorates and masters. Meanwhile, as the founder of Hubei Key Laboratory of Environmental and Disaster Monitoring and Assessment, he also set up the team of experts in the field of wetlands who still continues to move forward on the path of wetland research and wetland conservation.

5.2 Li Chang'an, a "Jingchu model" and "professor of mountains and rivers" committing to the conservation of wetlands of the Yangtze River

Li Chang'an is a professor and doctoral supervisor of China University of Geosciences (Wuhan). In his 40 years of work, he has devoted himself to teaching and research, and is particularly fond of wetland conservation.

As a scientist, he has been committed to research on the formation and evolution of the Yangtze River, and ecological and environmental conservation and sustainable development of the river basin. As a teacher, he is also actively involved in the popularization of science and technology for the ecological and environmental conservation of the Yangtze River basin; he takes advantage of his positions as a deputy to the National People's Congress (NPC) and a member of the Chinese People's Political Consultative Conference (CPPCC), he actively advises and advocates for the conservation of wetlands and the Yangtze River. His footprints have covered almost all the mountains and rivers of the Yangtze River basin, and he is known as the "professor of mountains and rivers". He has been awarded as "Wuhan model" and "Jingchu model".

After graduating from China University of Geosciences (Wuhan), Li stayed on as a teacher, mainly teaching and researching geomorphology and quaternary geology, and in 1998, a major flood in the Yangtze River basin shifted his research interests to the Yangtze River. Li Chang'an, who was conducting research in the northwest, received a call from his teacher, academician Yin Hongfu, "who wanted me to hurry back to Wuhan and provide scientific solutions for flood control of the Yangtze River through professional research from a geological perspective."

Li Changan came back and immediately went to investigate along the Yangtze River, "Standing on the dike, looking at the flood danger of the Yangtze River, I was very shaken inside. This is another natural disaster that left a deep impression on me after the Tangshan earthquake. Floods are too dangerous!" So he carried out several years of research on flood prevention and mitigation in the Yangtze River basin.

Then he discovered that the "age of the Yangtze River" had been controversial, and he focused his research on the formation and evolution of the Yangtze River. To this end, he spent nearly 10 years conducting research on the Yangtze River, walking almost all the main tributaries of the Yangtze River.

After years of scientific investigation, the ecological and environmental problems of the Yangtze River basin caught Li Chang'an's attention. "The Yangtze River is simply too important. Its population and GDP share levels are tremendous; its biodiversity has an important status in the international great river basin, and it is an important ecological and environmental barrier for China." During his extensive investigations, Li Chang'an realized the greatness of the Yangtze River and also discovered the health problems of this mother river. He also shifted the focus of his research to the environmental conservation of the Yangtze River, and has made outstanding achievements in the formation and evolution of the Yangtze River, flood prevention and mitigation, environmental conservation and sustainable development. He has won seven provincial and ministerial awards for scientific and technological achievements, as well as the title of "National Outstanding Geographical Scientist".

In 2007, the lake environment in Wuhan caught Li Chang'an's attention. "The lakes of this city play a huge role in flood prevention and water storage, ecological and environmental conservation and construction of ecological civilization."

For many years, Li Chang'an was either on his way to investigate or on his way to class. Li Changan drove around the lakes with his students every weekends. Before each survey, he first looked at the satellite remote sensing images, selected the key fields such as places with lake filling and shorelines with possible pollution hazards. He has basically visited all the major lakes in the three towns of Wuhan. In 2013, when he did an urban geological survey of Wuhan, he found that there was a pressing need to protect the mountains of the city, many of which had been destroyed by quarrying and mining, leaving the rock and soil exposed. He personally researched more than a hundred large and small mountains in Wuhan and put forward the proposal of "strengthening mountain conservation and improving ecological environment", which effectively promoted the mountain conservation and re-greening project in Wuhan.

The conservation of rivers and wetlands has always been on Li Chang'an's mind. In order to write the proposal of "Ecological and Environmental Conservation of Liangzi Lake", he approached the Liangzi Lake nearly ten times to conduct surveys and called for the establishment of a special "Liangzi Lake Ecological Special Zone" to protect the lake through administrative and legislative means. The in-depth series of "Report on Liangzi Lake" directly promoted the introduction of the "Plan of Ecological and Environmental Conservation of Liangzi Lake".

During his research on the Yangtze River basin, he found that the Yangtze River Basin is economically developed and populous, and is the most densely populated large river basin in the world. The conservation of the Yangtze River requires technology and, more importantly, the conscious conservation of all people. So Li Chang'an devoted a lot of time to promoting science popularization of conservation of the Yangtze River.

In the past ten years, Li Chang'an has spent his spare time in schools, institutions and communities, giving nearly 400 lectures of science popularization on the conservation of the Yangtze River, wetlands and mountains. In addition to his own science popularization of environmental conservation, he also uses his position as Chief Science Popularization Expert on Subject of Quaternary Geology of the China Association for Science and Technology, Head of Expert Panel for Science Popularization of Geomorphology and Quaternary Geology of Geological Society of China, Director of Hubei Science and Technology Communication Society and Director of Hubei Youth Science and Technology Association to lead and organize science and technology workers and social resources to widely carry out the science popularization of environmental conservation.

The Expert Panel for Science Popularization of Geomorphology and Quaternary Geology Li Chang'an leads has been awarded the National Advanced Science Popularization Team several times. His outstanding contribution to science popularization of environmental conservation has earned him the titles of "Top Ten Outstanding Figures in Science and Technology Communication in Hubei Province" and "Advanced Individual in National Science Popularization", and he was selected as a candidate for the Fourth National "Mother River Conservation Award". "He was selected as a candidate for the 4th National Mother River Conservation Award. His "universal view" and "scientific view" of environmental conservation of the Yangtze River are written into the compulsory education textbook "Middle School Politics" and the university textbook "Introduction to Environmental Science" respectively. In his words, "Popular science education can cultivate the scientific spirit of all people and cultivate a scientific way of thinking, which is very important for the progress and development of the nation."

While undertaking heavy tasks in teaching and research, Li Chang'an actively engages himself in social services. During his ten-year tenure as vice chairman of the Wuhan Committee of the China Association for Promoting Democracy, he was mainly responsible for political participation and Discussion. He took the lead in leading members to conduct a lot of research on the conservation of wetlands in rivers and lakes in Wuhan, resulting in a series of research reports and proposals and suggestions, making ecological and environmental conservation a highlight of the political participation and discussion of the Wuhan Committee. After he left his post, Wuhan Committee set up the "Li Chang'an Studio", which was featured in the *CPPCC Daily*.

When he was a CPPCC member, a member of the Hubei Provincial Committee of the CPPCC, a deputy to the Wuhan Municipal People's Congress and a counselor of the Wuhan Municipal Government, Li Chang'an conducted in-depth research with scientific attitude and actively wrote proposals, suggestions and consultation reports amounting to more than 80 about the well-coordinated environmental conservation of the Yangtze River, ecological and environmental conservation of rivers, lakes and wetlands in Wuhan city and Hubei Province, playing an important role in the construction of regional environmental conservation and ecological civilization and effectively promoting the conservation of lakes and wetlands in Wuhan and Hubei, the conservation and rational use of south-to-north water diversion and water resources, as well as the establishment of "registration system of national carbon trading" in Wuhan. Sixteen proposals have been approved by major leaders at the provincial level or above.

For his outstanding contributions to ecological and environmental conservation, Li Chang'an has been awarded the second prize of National Achievement in Political Participation and Discussion, Outstanding NPC Deputy of Wuhan, Outstanding Contribution Award of Wuhan Government Counsellor, Advanced Individual of Environmental Conservation of Hubei Province, and Advanced Individual of Lake Conservation of Wuhan City by the Central Committee of the China Association for Promoting Democracy. He received four awards of Excellent Research Achievement Award of Hubei Provincial Committee and Hubei Provincial Award of Development Research. The achievements of his political participation have been reported by China's important media such as *People's Daily*, *Guangming Daily*, *Chinese People's Political Consultative Conference Daily*, *China Environment News*, *Hubei Daily*, and *Yangtze River Daily*.

Currently, retired Professor Li Chang'an is still active in the field of environmental conservation. In the past two years, he has conducted a systematic survey of the Yangtze River basin in Wuhan city and Hubei province, calling for the conservation and management of small basins in the well-coordinated environmental conservation of the Yangtze River. He proposes the governance model of "mountains, rivers, forests, farmlands, lakes, and grasslands + farmers" and suggests the "agriculture, countryside + ecology" as the characteristic of a "dreamland of water town" to create a rural revitalization model with the integration of agriculture, tourism, culture, education and ecology. This year, Professor Li Chang'an planned and organized a scientific research on the Yangtze River source for college students under the theme of "well-coordinated environmental conservation of the Yangtze River".

For the sake of ecological conservation of rivers and wetlands, Professor Li Chang'an has been working hard to press ahead. Over the years, he has been dedicating all his time, knowledge and wisdom to the construction of a beautiful China and a beautiful Hubei. In the future, he will still "pursue in both directions" in scientific research and popularization of ecological and environmental conservation, in order to show his love for his family and country and responsibility of intellectuals.

5.3 Hubei Wetland Conservation Foundation, China's first of its kind

Hubei Wetland Conservation Foundation was born in Hubei on July 8, 2008, which is the first wetland conservation foundation established in China.

On June 10, 2005, the Ramsar Burear/Secretariat in Switzerland announced that the triennial Ramsar Wetland Scientific Award, the highest international prize for scientific research of wetlands, was awarded to Cai Shuming, a Chinese scientist for his outstanding contributions to scientific research and government policy-making on conservation and rational use of wetlands. He was also invited to participate in the 9th Conference of the Contracting Parties to the Ramsar Convention in Uganda. Cai Shuming, then a member of the National Committee of the Chinese People's Political Consultative Conference (CPPCC) and Vice Chairman of the CPPCC in Hubei Province, is a renowned wetland scientist in China, and his research results on wetland conservation and environmental change in Hubei Province have received special attention both at home and abroad. In order to further promote the conservation and management of wetlands in Hubei Province, Cai Shuming, as the recipient of the Ramsar Wetland Scientific Award, donated all the prize money he received and actively raised funds to establish the Hubei Wetland Conservation Foundation, which was strongly supported by the Provincial Government of Hubei Province, National Forestry Administration and the Central Committee of the China Association for Promoting Democracy. Mr. Peter Bridgewater, the then Secretary General of the Ramsar Bureau, who visited Hubei in 2007, highly appreciated this initiative.

The purpose of the Wetland Conservation Foundation is, guided by Xi Jingping's thought of ecological civilation, to promote the conservation and sustainable use of wetland resources in Hubei Province, to promote the construction of ecological civilization, to encourage and support the whole society to pay attention to the conservation and sustainable use of wetland resources, to promote the harmony between human and nature, and to contribute to the sustainable economic and social development of Hubei Province.

Since its establishment, the Hubei Wetland Conservation Foundation, under the guidance of the Hubei Provincial Forestry Administration and the Provincial Department of Civil Affairs, and with the strong support of the governing units and all directors, has fully played its role, actively increasing public awareness of wetland conservation, expert technical guidance on conservation and restoration projects and sustainable use of wetlands to promote harmony between human and nature and make contributions to the realization of high-quality green development of economy and society in Hubei. Hubei Wetland Conservation Foundation has also received support and sponsorship from many domestic enterprises that are keen on environmental conservation and public welfare. The entrepreneurs have taken practical action to add to the cause of environmental conservation of wetlands in Hubei, while wetland conservation has also formed the environmental conservation culture of these enterprises.

Hu Xinghuan is the current director of Hubei Wetland Conservation Foundation, while his predecessors are Cai Shuming, Song Defu, and Hou Kaiju.

5.3.1 Organize and carry out wetland-related popularization of law and activities of ecology and public welfare

(1) The Foundation organizes 1+n training activities with Wetland Conservation Law of the People's Republic of China as the main content. The experts are organized in a combination of centralized and divided form to carry out training on the *Wetlands Conservation Law*, related policies about wetland conservation and restoration, and project management.

(2) According to the resolution of the 68th United Nations General Assembly, March 3 of each year is the World Wildlife Day. The Wetland Foundation, the Hubei Provincial Forestry Administration and the Provincial Wildlife Conservation Association organize a large-scale publicity campaign for World Wildlife Day in Hubei Province every year.

(3) In accordance with the *Wildlife Conservation Law of the People's Republic of China*, the Foundation co-organizes the launching ceremony of the Hubei Bird Loving Week with the Hubei Provincial Forestry Administration and the Provincial Wildlife Conservation Association. Every November, the Foundation participates in the Month of Wildlife Conservation Publicity organized by the Hubei Provincial Forestry Administration and the Provincial Wildlife Conservation Association, and also holds a bird-watching competition with the Hubei Provincial Forestry Administration, the Provincial Wildlife Conservation Association, relevant wetland parks and nature reserves for employees with forestry system and volunteers. Through the activities, the Foundation deepens the knowledge of employees with wildlife conservation and wetland conservation authorities about wetland birds, improves the business level, and lays a solid foundation for wetland conservation.

(4) On February 2nd, the World Wetland Day every year, the Foundation, together with the Provincial Forestry Administration, relevant wetland parks, environmental organizations and other units, holds a publicity campaign for World Wetland Day. In the publicity activities, the primary and secondary schools that do well on wetland conservation education will be commended, and awarded the honorary title of Demonstration School of Wetland Conservation Education of Hubei Province. The following schools have been awarded the title: Wuhan Huazhongli Primary School, Wuhan East Lake Ecological Tourism Scenic Area Overseas Chinese Town Primary School, Wuhan Daxing Road Primary School, Wuhan Twenty-third Junior High School, Honghu First Primary School, Jingmen Zhanghe Town Central Primary School, Jingmen Shayang County Zengji Town Central Primary School, Xianning Xianan District Guanbu High School, Shiyan Dongfeng 22 Primary School, Nanzhang County Donggong Town Wanquan Primary School, Songzi City Weishui Town Dayanzui Primary School and education bases of comprehensive practice. In order to encourage more primary and

secondary school teachers to participate in wetland conservation education, the Wetland Conservation Foundation will award the title of Outstanding Teacher of Wetland Conservation Education in Hubei Province to teachers with outstanding performance in wetland conservation education as an encouragement.

(5) The selection activities of Hubei Wetland Conservation Award is jointly initiated by the Hubei Provincial Forestry Administration, Hubei Wetland Conservation Foundation, Hubei Daily Media Group, Hubei Provincial Media Group, WWF and other units the first time in the province, aimed at promoting ecological civilization, recognizing and rewarding individuals and groups who have made significant contributions to the cause of wetland conservation in Hubei and gained outstanding achievements.

(6) Promoting the formation of the Hanjiang River (Xiangyang section in Hubei) Wetland Park Alliance. The Foundation strives to advance the conservation and restoration of wetlands in the middle and lower reaches of the Hanjiang River, information sharing, upgrading and joint prevention and treatment.

5.3.2 Carry out wetland popularization and education and training activities to improve conservation awareness and management capacity

5.3.2.1 Carry out the photography contest of Ecological Hubei, Beautiful Wetlands

In order to welcome the 14th Conference of the Contracting Parties to the Ramsar Convention to be held in Wuhan in 2022, to better publicize the achievements of wetland conservation and restoration in Hubei Province, and to raise people's awareness of wetland conservation, Hubei Wetland Conservation Foundation has carried out a series of activities such as cooperating with Hubei Provincial Media Group to film the TV documentary The Great Beauty of Wetlands in Hubei and jointly carrying out the photography contest of Ecological Hubei, Beautiful Wetlands with Xinhuanet.

5.3.2.2 Have training courses of wetland conservation education and nature reserve management

Hubei Wetland Conservation Foundation has held a number of training courses of wetland conservation education for teachers in the whole province, and one course of construction and management of national wetland parks in the whole province. The 106 wetland nature reserves (wetland parks) and 304 principals, directors, publicity cadres and key teachers with 75 primary and secondary schools attended the training course. During the training course, experts and training instructors of nature experience were hired to teach. Training courses are held to continuously improve the knowledge of the managerial staff with each wetland reserve, wetland park, and wetlands about wetland conservation, techniques of wetland conservation, and management level of wetland conservation. Most of the schools have transformed the training results into teaching practice.

5.3.3 Carry out wetland conservation and restoration, resource monitoring and expert technical support

In recent years, through in-depth surveys and field visits to gain more details, and then inviting directors and experts to form an inspection team to organize consultation, the Foundation has determined to fund public welfare projects for seven wetland parks and nature reserves in the province.

(1) ecological monitoring project of Shennongjia Dajiu Lake National Wetland Park. The wetland monitoring of this wetland park has better basic conditions and greater social influence, and the project selection is unique, exemplary and scientific in the province.

(2) monitoring project of Xianning Xiangyang Lake National Wetland Park. The Xiangyang Lake is the southern half of the Futou Lake, the fourth largest lake in our province, and is an important ecological node in the Jianghan plain area. In recent years, a series of projects such as returning land for farming and fishing to lakes have been carried out, significantly improving the water quality of the wetlands and gradually restoring wetland biodiversity.

(3) ecological monitoring project of Hubei Fanwan Lake National Wetland Park. Fanwan Lake National Wetland Park is a typical shallow lake wetland in the lake groups of Jianghan plain, which is connected to the Hanjiang River at the upper part and to the Yangtze River at the lower part, and is an important ecological area for building a wetland security system in the Jianghan plain and even in the middle and lower reaches of the Yangtze River.

(4) bird survey project of wetlands of international importance in Wanghu Lake. The wetlands of international importance in Wanghu Lake are located in the migratory pathway of East Asia – Australia migratory birds with unique natural resources, superior geographic location, rich biodiversity, and advantageous position of bird species and population. They have a very important international conservation significance. In addition to the detailed background survey and solid groundwork, it is necessary to develop cooperation in wetland conservation.

(5) conservation and monitoring project of Xiangyang Nanzhang County *Aix galericulata* and Habitat. The project is located in Changji village, Donggong town, Nanzhang county, Xiangyang, which was awarded the Town of *Aix galericulata* in China by China Wildlife Conservation Association in 2014 and is one of the three major concentrated habitats of *Aix galericulata* in China. Aix galericulata has a very high ecological status and important species

value in the world of biological species conservation. I will support and fund the implementation of conservation and scientific research monitoring, science popularization education, resource conservation, and wintering food cultivation of Aix galericulata and habitat.

(6) treatment project of production and domestic sewage biodegradation of Jingmen Zhanghe RiverNational Wetland Park. The project is located in the Liji Island of Zhanghe River National Wetland Park in Jingmen city. The implementation of the project can prevent the water quality pollution of the reservoir area from the direct discharge of production and domestic sewage on the island, and provide models and examples of production and domestic sewage treatment for the majority of rural areas. For this reason I will support the Zhanghe National Wetland Park to collect and concentrate domestic and production sewage into artificial ponds with artificial wetland treatment model, and make the discharge pollution-free after three levels of degradation treatment of aquatic plants or microbe of artificial ponds.

Hubei Wetland Foundation and funded units jointly carry out monitoring of wetland resource and migratory birds in order to participate in the construction and supervision of implementation and create a demonstration project of fine wetland ecological conservation with certain influence in Hubei Province. By funding the above basic research and public welfare projects of wetlands, the Wetland Foundation aims to publicize the good cases of wetland conservation and promote the good technical model as a demonstration case. The implementation of these projects has achieved good socio-economic and environmental benefits; at the same time, it has also expanded the social influence of Hubei Wetland Conservation Foundation, representing by a clear sign plate of Hubei Wetland Conservation Foundation in each project site.

5.3.4 Strengthen its own construction and social participation to enhance the influence of the Foundation

(1) Publishing a *General Knowledge Book on Wetland Conservation*. In order to popularize the knowledge of wetland conservation to the public, and also to come up with the Foundation's own promotional materials, the Foundation organized the preparation of the *General Knowledge Book on Wetland Conservation*. The book includes contents such as "What is a wetland", "Main types of wetlands", "Main functions of wetlands", "Threats to wetlands" and "How to protect wetlands". As the basic promotional material of the Foundation, the book is planned to be distributed to the public at the wetland promotion activities, and is also intended to be distributed at the International Wetland Conference to be held in Wuhan next year to showcase the promotional work carried out by the Provincial Wetland Conservation Foundation.

(2) In order to make full use of the Internet to disseminate widely, the Foundation has actively turned to the Internet and set up its official account "Jingchu Wetland Conservation". The official account is divided into three major sections: one introduces the Foundation; one introduces the work and activities carried out by the Foundation; and the other popularizes the knowledge of wetland conservation. The construction of the official account has been completed, and it is now running on a trial basis to be launched to the public at the right time. The official account will also actively seek cooperation with institutions, units and individuals who are interested in protecting wetlands, and jointly operate the account to make "Jingchu Wetland Conservation" a publicity window of the Provincial Wetland Foundation.

5.3.5 Organize social welfare fundraising and manage wetland conservation fund

On the basis of further social awareness of the importance of wetland conservation, the Foundation actively mobilizes more entrepreneurs and wetland conservation enthusiasts to donate, raising more funds for the development of wetland conservation in Hubei Province. At the same time, the Foundation strictly follows the *Charter of Hubei Wetland Conservation Foundation* to make sure the wetland conservation fund is well managed and well used.

5.4 WWF, a NGO dedicated to biodiversity conservation

5.4.1 WWF and biodiversity conservation

WWF (World Wide Fund For Nature) is the world's largest independent environmental NGO with a global reputation. It dedicates itself to the conservation of the world's biodiversity and its habitat, and strives to reduce the impact of humans on these organisms and their environment.

WWF's work in China began in 1980 with the conservation of giant pandas and their habitats, and was the first international NGO to be invited by the Chinese government to carry out conservation work in China. WWF's mission is to stop the degradation of the Earth's natural environment and to create a better future where people live in harmony with nature. To this end, we are committed to protecting the world's biodiversity, ensuring the sustainable use of renewable natural resources, and promoting actions to reduce pollution and wasteful consumption.

5.4.2 WWF and wetland conservation in Hubei

WWF is the first international environmental organization to work on nature conservation in China. Since 1994, WWF has been carrying out projects of wetland conservation and sustainable use in China, and assisting the National Forestry Administration and 17 ministries in developing and publishing the Action Plan of China Wetland Conservation.

Since August 2002, WWF has been working on the effective conservation of wetland ecosystems in the middle and lower reaches of the Yangtze River, especially the ecosystems of barrier lakes, and the sustainable use of wetland resources. In Hubei, it has successively carried out the conservation of *Neophocaena asiaeorientalis*, environmental flow, transformation of fishery market, "rebuilding the web of life of the Yangtze River", "walking with climate partners", "small species for wildlife conservation", creation of international wetland cities and wetlands of international importance, construction of national wetland parks, and preparation for the Conference of the Contracting Parties to the Ramsar Convention.

Through field demonstration, cooperative network building, policy advocacy and publicity, and environmental education, the project has carried out fruitful work in promoting community-based conservation and rational use of wetland resources, rebuilding river-lake linkages, building a network for the conservation of wetlands of the Yangtze River and Cetacea, integrated basin management, and demonstration environment-climate friendly agriculture models, providing governments and local communities at all levels many effective solutions and practical examples.

In 2004, the Hubei Provincial Forestry Administration and WWF co-established the Hubei Wetland Reserve Network, exploring ways to improve the effective management of wetlands through the creation of wetland reserve network. Based on this model of cooperation, the former National Forestry Administration and the Provincial and Municipal Governments of the Yangtze River basin jointly supported the creation of the Yangtze River Wetland Network, a network platform at the basin level. The network aims to promote the conservation and sustainable use of wetlands, to jointly address the impact of global climate change on regional sustainable development, and to improve the capacity of wetland conservation and management in the basin.

5.5 Other social organizations and figures in wetland conservation

5.5.1 Hubei Wildlife Conservation Association

Hubei Wildlife Conservation Association has done a lot of work in wildlife science popularization, ecological ethics education for minors, and ecological culture dissemination, which has produced a good social impact and made positive contributions to strengthening biodiversity conservation and promoting construction of ecological civilization.

The Association mainly focuses on strengthening popularization of wildlife conservation, carrying out ecological ethics education for minors, spreading ecological culture, and assisting government departments to carry out wildlife conservation work. The Association has sponsored four World Wildlife Day, four provincial Bird-Loving Week, four provincial publicity activities of wildlife conservation, published and distributed four books on ecological ethics education for minors, compiled and printed a compilation of experience in ecological ethics education for minors, carried out various forms of fly conservation activities, rescuing more than 80 species of birds and more than 1,200 birds, held fly conservation actions of migratory birds and more than 100 popularization activities into schools, communities, villages and families, and more than 10,000 copies of publicity materials have been distributed. The Association's work has been solidly promoted and won many honors. Three members were awarded advanced individuals by the National Forestry Administration for protecting forests and wildlife resources, and eight members were awarded the Subaru Ecological Conservation Award by the China Wildlife Conservation Association.

5.5.2 Wuhan Bird Watching Society, home of volunteers

Wuhan Bird Watching Society is a social welfare group dedicated to research of wild birds, conservation and ecological environmental conservation. It was officially registered with the Wuhan Bureau of Civil Affairs on March 30, 2017, and the business unit in charge is the Wuhan Landscape and Forestry Administration.

Over the years, the Society has been insisting on carrying out nature education, bird monitoring, bird research, bird rescue and care, and conservation of bird habitats in Wuhan by means of volunteer services. "Together, we make fun meaningful." is the consensus of all members, and promoting the harmonious coexistence of human and nature is the goal of the Society.

In July 2016, the Society mobilized volunteers, mainly its members, to carry out a public welfare project of bird monitoring in key areas of Wuhan, gathering bird-watching enthusiasts to conduct regular bird observation, scientific research and conservation of birds and their habitats. The main objectives of the project are to study the classification and distribution of birds in Wuhan, monitor the dynamic changes of birds in key areas, and discover factors that threaten the survival of birds. By the end of 2021, with the efforts of volunteers, the 70 key regional bird monitoring sites were established throughout the three towns of Wuhan, including wetlands of international importance of Chenhu Lake in Caidian and provincial and municipal wetland nature reserves of the Zhangdu Lake in Xinzhou, the Shangshe Lake in Jiangxia, the Caohu Lake in Huangpi, and the Wuhu Lake in Hannan, national wetland parks of the East Lake in Wuhan, the Houguan Lake in Caidian, the Canglong Island in Jiangxia, the Anshan Mountain in Jiangxia, the East-West Lake and Jinyin Lake, and the Dugong Lake, as well as urban forest parks of the Ma'an Mountain in Wuhan, the Bafen Mountain in Jiangxia, and the Mulan Mountain in Huangpi, covering various habitat types such as wetlands, forests, farmlands, urban green areas and residential areas.

A total of 9,288 monitoring reports have been received in five years, 64 monthly bird monitoring reports have been published to the society, and annual bird monitoring reports have been compiled for six consecutive years. By collecting, counting, summarizing and publishing the classification, distribution and dynamic changes of birds in Wuhan, we have accumulated a large amount of scientific data for ecological and environmental conservation in Wuhan and made positive contributions to the construction of ecological civilization in Wuhan and even the well-coordinated environmental conservation of the Yangtze River.

With the mobilization and encouragement of the Society, some members have been continuing to carry out actions for bird rescue, and in the past five years, a total of 2,348 bird rescues have been carried out, rescuing 134 species of wild birds of about 17 orders and 41 families, totaling 3,564 birds. Among them, there are 23 species of national Class II key protected birds and 30 species of provincial key protected birds.

The Society has been insisting on carrying out public service of nature education and popularization of knowledge about bird conservation, carrying out 128 free activities of nature education with the theme of wetlands and migratory bird conservation in the East Lake Greenway, wetland parks, forest parks and communities, going into primary and secondary schools and holding offline lectures, covering about 40,000 attendances.

The Society compiled and published the Wuhan Bird Atlas, which summarizes the results of 20 years of bird observation in Wuhan, comprehensively introduces the 419 species of wild birds recorded and confirmed as of 2020, and officially publishes the *Wuhan Bird List* for the first time, delivering scientifically important research results of bird resources in Wuhan, and providing popular and detailed learning tools for bird observation and nature education for the public. The *Wuhan Bird Atlas* won the 2021 Outstanding Science Book Award in Wuhan City. The Society also edited and published Birds Outside the Window: Wuhan Indoor Report on Bird Watching, which is a beautiful product of the deep communication between human and nature during the special period of Wuhan's fight against Covid-19 in early 2020, and a power transmission of the harmonious coexistence between human and nature. Birds *Outside the Window: Wuhan Indoor Report on Bird Watching* won the Reader's Recommendation Award of the 2nd Pingshan Nature Museum Book Award.

5.5.3 Wetland schools allow ecological education in schools

5.5.3.1 Wuhan Huazhongli Primary School: bird-loving and lake-conserving for 33 years, setting a benchmark for wetland education

Adhering to the school philosophy of "adaptation and coexistence, suitability and growth", Huazhongli Primary School has started its journey of wetland education since 1989.

Beginning: bird conservation as a medium to conduct special education. The school started its "bird-loving and bird-care" education in 1989, and established the first bird-loving club for students in China. It combines bird-loving education with subject curriculum, moral education activities and traditional culture, and has developed the school-based teaching material Bird Language and Tang Poetry. Every Bird-Loving Week, the school carries out a variety of activities to express their love for birds through talking, dancing, singing, painting, writing, watching, making bird nests, and advocacy of bird conservation, and promotes the importance of bird conservation to the public. The students and teachers have not only been to the wetland parks in Wuhan city for bird-watching and bird-conserving, but also expanded their coverage to the surrounding Chenhu Lake and Jingshan Mountain.

Base: wetlands on campus for regular education. Years of "bird-loving and bird-care" action let the students and teachers of Huazhongli Primary School understand a truth: simply loving birds and conserving birds is not enough; to protect birds, we must protect wetlands, habitats of birds. In 2006, in accordance with the idea of "overall planning, gradual investment and gradual construction", the school has invested more than 1 million yuan to establish the first wetland ecology education museum of elementary school and rooftop wetland garden, and developed the school-based curriculum Wetlands - the cradle of life, allowing its wetland education to become regular. In 2009, the school was awarded the title of "Wetland Experimental School" by China Office of Wetland International. It hosted the 13th World Lakes Conference for Chinese, Japanese and Korean youth for its special wetland education, once again becoming the focus of attention for its bird-loving and lake-conserving characteristics.

Extension: off-campus investigation for in-depth practice. In 2015, taking into account the characteristics of Wuhan as the city of 100 lakes, the school developed the school-based teaching materials Beautiful Wuhan, Colorful Lakes by selecting 14 typical lakes from 166 lakes in Wuhan, in order to further enrich the connotation of wetland education. The materials guide students to pay attention to local lakes and wetlands in Wuhan, to feel the changes of the lakes in their hometown, to inspire love for the lakes, and to develop actions to protect the lakes, from the aspects of the lake style, the history of the lake, the famous places of the lake, the present of the lake, the poems and songs of the lake, and the impressions of the lake tour. From 2017, the school has been planning a series of investigation and practice activities such as "love my a hundred lakes - urban green lungs", "tour the lake to enjoy the lotus" and "lakes at my doorstep" with the theme of "care about wetlands, love my a hundred lakes, and be a little messenger to conserve the lakes" in the summer vacation, organizing students to go to the major lakes in Wuhan to conduct wetland investigation, interview citizens, and write science popularization records.... The themed activities are conducted from in-school to out-of-school, and then from out-of-school to in-school, from knowledge learning to field experience, and then to conservation action, promoting wetland education at multiple levels and angles. In 2018, the school was awarded the Demonstration School of Wetland Conservation in Hubei Province by the Hubei Wetland Foundation and the first batch of Youth Environmental Volunteer Action Pilot School for "Beautiful China, I am an Actor - Small River Chief, Small Lake Chief" by the Publicity and Education Center of the Ministry of Ecology and Environment..

Perseverance: pressing forward poetically to set up the benchmark of lake conservation. From the establishment of the first bird-loving and bird-conserving club for elementary school students to the establishment of the first wetland ecology education museum of elementary school, to being amony the first batch of international wetland schools, the first batch of Youth Environmental Volunteer Action Pilot School for "Beautiful China, I am an Actor - Small River Chief, Small Lake Chief", the first batch of international ecological schools, the school has been committed to "bird-loving and lake-conserving" for 33 years. From bird conservation to wetland education, from in-school to out-of-school, and then from regular popularization and education to field practice, "green" has become the colorful background of the school, and "let birds have a home" has become the solemn declaration of all the teachers and students. The "bird-loving and lake-conserving" actions such as bird watching in spring, lake research in summer, nature experience in autumn, and migratory bird monitoring in winter have become a big miracle of ecological civilization, making Huazhongli Primary School the only school in central China that has been a benchmark for bird and lake conservation for 33 years.

5.5.3.2 Overseas Chinese Town Primary School Affiliated to Central China Normal University: embracing green East Lake for a colorful childhood

The Overseas Chinese Town (OCT) Primary School Affiliated to Central China Normal University is located on the north shore of the East Lake, a 5A national ecological tourism scenic area. It relies on the unique ecological environment of the lake, attaches great importance to the construction of ecological civilization, focuses on cultivating environmental awareness among teachers and students, so that the seeds of ecological civilization are deeply rooted in children's hearts, internalized in their minds and externalized in their actions.

Taking the "ecological East Lake" as the starting point, the school establishes the "ecological civilization education" as the characteristics of school, and builds the "Education Ecosystem of OCT Primary School". The shool pools knowledge of the ecosystem of the lake, and through quizzes, games, drawings and practical activities guides children to learn about the East Lake ecosystem in an inquiry-based and in-depth manner, to form a new ecological view of nature and ecological morality, and to advocate the conservation of the ecological environment with real actions.

Research and development of teaching materials to deepen the concept of wetland ecological civilization. In accordance with the requirements of the new curriculum standards, OCT Primary School makes education of wetland ecological civilization a part of the whole process of teaching literature and language, mathematics, science, art and other subjects, fully explores the factors of low-carbon environmental education in the teaching materials, integrates the subjects, and establishes the awareness of ecological and environmental education. The school has prepared three school-based teaching materials Charming East Lake, Humanistic East Lake and Ecological East Lake; and it deepens the connotation of wetland conservation education by undertaking the research topics of the China Young Pioneers and Hubei Provincial Scienctific Planning on Education.

Building bases to enhance the awareness of ecological civilization of wetlands. First, the East Lake Ecological Civilization Education Museum in the school shifts the focus from on the big topic of "ecological civilization education", the big article to the "small" topic of "ecological East Lake", respectively displaying the achievements of urban green center of the East Lake, wetland conservation, water quality optimization, environmental treatment and biodiversity to inspire students to understand, love, protect, and make contributions to the East Lake. Through the sound and light technology, it provides a new world for teachers and students to communicate with the East Lake in a large view and close distance. Second, the school sets up the off-campus social practice base of the Huayang Lake where teachers and students learn specific scientific methods to protect the wetlands of the Great East Lake in order to see the big picture in the small wetland environment of the Huayang Lake.

Practice activities to create a home of wetland ecological civilization. First, the school carries out regular patrol of the lake, take photos and videos to capture the changes of the lake, take notes to record its beauty, conserving the wetlands of the East Lake in real actions. Second, the water quality of the East Lake is monitored. The causes of water quality changes will be investigated and analyzed every two weeks to ensure that the status of water resources and environment of the wetlands of the East Lake is monitored. Third, the students assume the role of small lake chiefs. Students have also developed the school-based teaching materials Huayang Lake, set up an eco-talking group, and carried out activities such as the environmental conservation forum of Sunshine Parents, the introduction of experts to the campus, visits to scientists, the Eco-Art Festival of "Embrace Green, Colorful Childhood", "Green LOVE" the sports meeting of low-carbon environmental conservation, and the World Environment Day.

5.5.3.3 Wuhan Daxing Road Primary School: a strong love for wetlands of the Hanjiang River

Wuhan Daxing Road Primary School, which has won the "Global 500" Youth Environment Award by United Nations Environment Programme, has been "green" for its environmental education for more than 20 years. But the school's "green" march has never stopped, instead, it always seeks innovation and breakthroughs. Since 2012, Daxing Road Primary School has been engage in wetland education and has continued to carry out ecological expeditions on the Hanjiang River. During this period, it once spent five years leading students to investigate the Hanjiang River, which was a ecological field activity with much characteristics and have become an indelible mark in the children's growth process and opened up a brand new path of ecological education.

The environmental education of the school has kept pace with the times. Taking advantage of its proximity to the confluence of the Yangtze River and Hanjiang River, and its geographical advantage of being within a stone's throw of the Longwang Temple in Hankou, the school turns its attention to the Hanjiang River, the "source of life" in Jingchu. In June 2012, Daxing Road Primary School set up the first off-campus practice base for water and environment education in the Longwang Temple in Hankou in the Hanjiang River basin, and started research on river wetlands.

In August 2015, with the theme of "Love of Hanjiang River, Destiny of Crested Ibis", the school dovetailed with the Sino-Japanese cooperation project on the construation of regional environement for harmonious coexistence of human and Crested Ibis, and carried out environmental practice activities for ecological and moral education based on the Hanjiang River inspection.

The results of the Hanjiang River inspection from Daxing Road Elementary School were widely noted and recognized. In 2014, the school received the Environmental Conservation Government Award of Hubei Province, and in 2016, the school was awarded the 7th Mother River Award by the Central Committee of the Communist Youth League of China and eight other central ministries and commissions, which provided unlimited motivation for the school to deepen its ecological inspection activities of the Hanjiang River.

The first phase of the school's five-year inspection plan is over, and the second phase will focus on the changes in water quality and wetland conservation of the Hanjiang River in the context of the development of the South-to-North Water Diversion and the Hanjiang River Economic Belt every five years. Today, these inspection activities have become a special brand of Daxing Road Primary School, setting an eye-catching benchmark for wetland conservation practices.

However, the journey of scientific investigation is far from over, and promoting the sustainable development of wetland education is still a long-term subject. In this regard, the exploration Daxing Road Primary School will never stop.

5.5.3.4 Wuhan No. 12 Middle School: imprinting on endangered fish species of the Yangtze River

In the showroom of Wuhan No. 12 Middle School, the black-and-white engravings clearly record the "identity information" of endangered fish species in the Yangtze River. These are the works of the project "Imprinting on Endangered Fish Species of the Yangtze River" led by Liu Na, a teacher of the school since 2018, in which the endangered fish species of the Yangtze River is rejuvenated and revitalized under the students' carving knives.

The engraving of endangered fish species in the Yangtze River originated from "Keeping the smile of the porpoise", a lecture conducted by the school. In the summer of 2018, students from eight schools in Qingdao and the "Light and Shadow Club" of Wuhan No. 12 Middle School followed the footsteps of this lecture and started a "journey to find porpoise". They visited the Aquatic Animal Museum of the Chinese Academy of Sciences, made a donation to the Wuhan Baiji Conservation Foundation, and tested 20 bags of mud samples collected from the *Neophocaena asiaeorientalis*'s habitat to find out the acidity, alkalinity, composition, whether it contains heavy metals, and to analyze the *Neophocaena asiaeorientalis*'s living environment.

The "Light and Shadow" Roots and Shoots Group of Wuhan No. 12 Middle School initiated a project entitled "Guarding the Yangtze River Elves: Imprinting on Endangered Fish Species". The students have made engravings of the "silent extinct" in the water, and have completed engravings of more than 100 species of the Yangtze River. Their goal is to create a "family engraving" of the fish species of the Yangtze River.

Since the launch of the "Guarding the Yangtze River Elves: Imprinting on Endangered Fish Species" project, the students have been bringing their elaborate engravings to different places and inviting people to create with them. They have been active at the Yangtze River Civilization Museum, Wuhan Science and Technology Museum, Hankou Linghu Lake Square, and Jiefang Park in Wuhan. In 2021, the students and teachers were invited to participate in the COP15 NGO Parallel Forum in Kunming, Yunnan Province, to speak out for the diversity preservation of fish species of the Yangtze River.

5.5.4 Managers and volunteers at the front line of wetland conservation

5.5.4.1 Feng Jiang with Caidian Chenhu Wetland Nature Reserve in Hubei

Advanced Individual of Bird Loving and Bird Conservation in Wuhan in 2010, Excellent Individual of Construction and Management of Nature Reserve in Hubei Province in 2013, Civilized Citizen of Wuhan in 2016, nomination of The Most Beautiful Wildlife Guardian in Hubei Province in 2017, nomination of The Most Beautiful Grassroots College Graduate in 2019, Model of Wuhan in 2019, Advanced Individual of Bird Loving and Bird Conservation in Hubei Province in 2021

5.5.4.2 Yang Tao with Shishou Milu National Nature Reserve in Hubei

He joined Shishou Milu National Nature Reserve in 2011 and is mainly engaged in the scientific research and monitoring of the reserve. As a college student graduated from a professional college, he can stand loneliness and bases his life firmly in the reserve. He studies diligently and works hard to do a good job in patrolling and monitoring. He has developed a monitoring program, established a biodiversity monitoring system in the reserve, and found out the "family background" of wild animals and plants in the reserve. He loves his career and is dedicated to rescuing young astray *Elaphurus davidianus*. With the help of experts and his colleagues, he has successfully rescued the injured *Elaphurus davidianus* and released them into the wild to their herd. He has made scientific suggestions to strengthen restoration and conservation of wetland habitats. He actively participates in the comprehensive scientific study and results summary, and also has independently written or co-written and published reports such as *Study on the Diversity of Birds in the Wetlands of the Tian-e-zhou Oxbow in Shishou, Hubei* and *Current Status and Measures of Epidemic Prevention and Control of Elaphurus davidianus in Shishou Milu Reserve*, providing suggestions for the conservation, restoration and management of the reserve.

5.5.4.3 Ding Zeliang with Yangtze River Tian'ezhou Baijitun National Nature Reserve in Hubei

He is currently responsible for the training, breeding and management of Yangtze *Neophocaena asiaeorientalis* reared in cages in the Tian-e-zhou oxbow, as well as the monitoring, patrolling and inspection of the environment, fishery resources and population of the porpoises around the Tian-e-zhou oxbow. Ding Zeliang was formerly a professional fisherman in the Tian-e-zhou oxbow, living by the Yangtze River since childhood. He has been employed by the Yangtze River Tian'ezhou Baijitun National Nature Reserve since 2008, and as the most important technical force of fishermen has participated in various surveys of the Yangtze *Neophocaena asiaeorientalis* population organized by the Institute of Hydrobiology, Chinese Academy of

Sciences, and the ex-situ conservation work of the Yangtze *Neophocaena asiaeorientalis*.

5.5.4.4 Yan Jun with Wuhan Bird Watching Society

Yan Jun is a bird conservationist and expert in nature education in Wuhan. He is one of the founders of Wuhan Bird Watching Sociey, and is currently the executive director of Hubei Wildlife Conservation Association, president of Wuhan Birdwatching Society, and executive director of Nature Education Branch of Wuhan Park Association.

For 15 years, he has been insisting on observing, photographing and studying wild birds in Wuhan and the surrounding areas, promoting bird watching activities, and carrying out bird survey and monitoring, bird conservation and habitat conservation under the guidance of relevant government departments. He has published more than ten papers on bird research including Preliminary Study on the Distribution of Aythya baeri in Huangpi District, Wuhan, led the compilation and publication of the Annual Report on Bird Monitoring in Key Areas of Wuhan for six consecutive years. He is also the editor-in-chief of Wuhan Bird Atlas and Bird Watching Manual, and main author of Birds Outside the Window: Wuhan Indoor Report on Bird Watching. In 2017, he was awarded the Advanced Individual of Bird Loving and Bird Conservation in Hubei, the Advanced Individual of the National "Volunteer Fly Conservation Action" in 2018, and The Most Beautiful Volunteer in Wuhan City in 2019.

5.5.4.5 Wang Dexin, a volunteer patroller in Yuan'an Juhe National Wetland Park in Hubei

As a volunteer patroller in Yuan'an Juhe National Wetland Park in Hubei, he is one of the guardians of the wetland park, silently guarding the green home. Wang Dexin's home is next to the Juhe National Wetland Park, and he worked for the Wetland Park as one of the patrollers as public-service jobs from 2018 to 2021. In 2022 the Wetland Park cancelled the public-service jobs, then Wang Dexin has been volunteering to become a patroller, and popularizing science for the surrounding residents and visitors.

5.6 Scientific research support for wetland conservation in Hubei

Hubei, a great land fostering great men, is a province of science and education with strong educational resources: the universities in Hubei Province are among the top in quality and quantity, and it also has a large number of research institutes. Relying on its good scientific research platform of research institutes and universities, Hubei has built several laboratories and research centers related to wetland science, and provided a good platform basis for the development of scientific research related to wetland conservation and monitoring in Hubei Province.

Hubei Province actively participates in international and domestic cooperation and exchange of wetland conservation, carries out scientific monitoring and research on wetlands, and continuously improves its level of scientific research. The higher education institutions and scientific research units such as the Innovation Academy for Precision Measurement Science and Technology, Chinese Academy of Sciences, Institute of Hydrobiology, Chinese Academy of Sciences, Wuhan University, Wuhan Botanical Garden, Chinese Academy of Sciences, and China University of Geosciences have carried out ecological positioning observation and research of the welands in Honghu Lake, Liangzi Lake, Tian-e-zhou Oxbow, Chenhu Lake, Longgan Lake, and Shennongjia Dajiu Lake, and have done a lot of productive work and achieved fruitful scientific research results in the study of lakes and marsh wetlands, prevention and control of wetland pollution, conservation, research and scientific cooperation as well as support and exchange of rare aquatic animals such as *Lipotes vexillifer*, *Neophocaena asiaeorientalis*, and *Acipenser sinensis*.

5.6.1 Innovation Academy for Precision Measurement Science and Technology, Chinese Academy of Sciences

Since the establishment, Innovation Academy for Precision Measurement Science and Technology, Chinese Academy of Sciences and Hubei Key Laboratory of Environmental and Disaster Monitoring and Assessment have undertaken scientific research projects of national and local governments and achieved many scientific achievements such as the Ramsar Wetland Scientific Award. Published *Hubei Wetland Ecological Protection Research Series (10 volumes)*, a special fund project for academic works in Hubei Province, which is the first wetland protection research series in Hubei Province. The Laboratory applies remote sensing technology and positioning experimental methods to monitoring the evolution and dynamics of wetland ecosystems, and researches the degradation mechanism of damaged wetlands and the theory and technology of restoration and reconstruction based on the ecological process and environmental effects of wetland systems. We have a field scientific research station for wetland ecosystems in Honghu Lake of Chinese Academy of Sciences, a field scientific research station for wetland ecosystems in the Jianghan Plain of Hubei Province, and an engineering technology research center for surface source pollution in Hubei Province. Over the years, the Academy has continued to carry out a lot of scientific research work related to wetland conservation in Honghu Lake, Changhu Lake, Dajiu Lake, Liangzi Lake, Wanghu Lake, Longgan Lake, Chenhu Lake, Tangxun Lake, Baoan Lake, and the main tributaries of the Yangtze River and continued to make advice and suggestions, providing indispensable scientific support for the development of wetland conservation and the improvement of the conservation system in Hubei Province.

5.6.2 Institute of Hydrobiology, Chinese Academy of Sciences

Institute of Hydrobiology (IHB), Chinese Academy of Sciences is a comprehensive academic research institution specializing in the studies of life processes of inland aquatic organisms, ecological environment protection and utilization of biological resources. Facing the major strategic needs of the country, Institute of Hydrobiology focuses on fundamental, strategic and prospective scientific and technological issues in the field of life processes of inland aquatic organisms, ecological environment protection and utilization of biological resources, and plays a leading role in the field of water environment protection, freshwater fishery and microalgal biotechnology. It has co-established institutions and analytical testing center platforms such as the national field scientific observatory of lake ecosystem of the East Lake in Hubei, and the experimental station of Xiangxi River ecosystem in Three Gorges reservoir. Cao Wenxuan, an academician of Chinese Academy of Sciences, is the first advocate of the 10-year fishing ban on the Yangtze River, which has become an important action plan in the national strategy of "well-coordinated environmental conservation of the Yangtze River". Since the establishment of IHB, 200 achievements have been awarded, and some important technical innovations or technology integration have been successfully applied, which have played a significant role in promoting water environment conservation and sustainable development of fishery in Hubei Province and China.

5.6.3 Wuhan University and the national field scientific observatory and research station of lake ecosystem of Liangzi Lake

The national field ecological station of Liangzi Lake of Wuhan University takes the grass-type lakes and aquatic vegetation in the middle and lower reaches of the Yangtze River as the research objects to explore the mechanism of impact of natural disasters and environmental pollution on the lake ecosystem and to provide first-hand data and technical support for the ecological environment conservation of lakes and the restoration of damaged freshwater ecosystems in China. Wuhan University started to conduct long-term positioning observation of aquatic ecosystem and reconstruction experiment of degraded aquatic vegetation in the Liangzi Lake in 1992, and has successfully planted 200,000 mu of aquatic vegetation, providing basic guarantee and technical support for the improvement of water quality of the lake.

The national field scientific observatory and research station of lake ecosystem of Liangzi Lake in Hubei is the first national station that takes aquatic plants and ecological environment conservation of lakes as research objects. It serves the needs of national water pollution treatment and lake conservation in Hubei Province, strives to make its due contribution to the ecological environment conservation and restoration of the Liangzi Lake, and provide demonstration models for the treatment and ecological restoration of the increasingly serious lake eutrophication in the country.

5.6.4 Wuhan Botanical Garden, Chinese Academy of Sciences, China University of Geosciences (Wuhan) and Hubei Provincial Key Laboratory of Wetland Evolution and Eco-Restoration

Hubei Provincial Key Laboratory of Wetland Evolution and Eco-Restoration was jointly established in October 2008 by Wuhan Botanical Garden, Chinese Academy of Sciences and China University of Geosciences (Wuhan). The laboratory aims at the frontier of the discipline, emphasizing the interdisciplinarity of geography, geology, biology and ecology. Its research aims at the sustainable development of wetland ecosystems in Hubei Province as the overall goal, combining the structural and functional characteristics of wetland ecosystems in Hubei Province, the threatened situation, and the characteristics of biodiversity conservation, and carrying out relevant monitoring and research work including dynamic monitoring of wetland ecosystems, biological and ecological foundations, as well as the theoretical and methodological approaches for the restoration and reconstruction of damaged wetland ecosystems.

5.6.5 Hubei Academy of Forestry (Hubei Branch of Chinese Academy of Forestry)

Hubei Academy of Forestry was established in 1959, and is a provincial forestry comprehensive public-service research unit mainly for applied research and technology development. In 2011 Hubei branch of Chinese Academy of Forestry was established. The Academy has always insisted on the ecological and economic development strategy of the province and the development concept of "Ecological Hubei, Green Rising" to actively carry out scientific and technological research, promotion and services.

In the field of wetland conservation and restoration, it was approved by the National Forestry Administration in 2014 to set up the positioning research station of wetland ecosystem of Honghu Lake in Hubei, and in 2021 it was approved by the Provincial Forestry Administration to formally establish the Wetland Research Institute. In 2022, the National Forestry and Grassland Administration approved the establishment of the National Innovation Alliance for the Conservation and Restoration of Wetlands in the Middle Reaches of the Yangtze River. The Academy also carries out scientific research and technological innovation in wetland biodiversity, wetland ecosystem services, ecological restoration and reconstruction of wetlands, technology research of wetland conservation and management, wetland ecology and wetland carbon sink/source monitoring in the province.

5.6.6 Hubei Provincial General Station for Wildlife Conservation

Hubei Provincial General Station for Wildlife Conservation is responsible for the survey and monitoring of wildlife, nature reserves and wetland resources in the province, as well as the construction of the nature reserve network; it organized and implemented the sub-project "strengthening the wetland conservation system and conserve biodiversity in Hubei Province" of the project "mainstream of life - strengthening wetland conservation system and conserving biodiversity" funded by the Global Environment Foundation. By improving the provincial policy framework, optimizing the basin-level management system, and strengthening the capacity of nature reserves, the project has improved the effectiveness of the management of wetland conservation system in Hubei Province and protected globally important biodiversity and wetland ecosystems. And the station also timely summarizes the results of various projects, and have published *Wetlands in Hubei, Waterbird Resources and Conservation in Hubei Province, Monitoring Manual of Hubei Wetland Nature Reserves, Study on Biodiversity and Ecological Value Assessment of Wetlands of International Importance in Chenhu Lake, Hubei Wetland Biodiversity Conservation and Construction of Nature Reserves, and Traveling with Wetlands. Traveling with Wetlands* won Excellent Science Popularization Works in Hubei Province in 2019.

In addition, scientific research institutions and colleges and universities such as the Changjiang River Scientific Research Institute of Yangtze River Water Resources Commission, Changjiang Water Resources Protection Institute of Yangtze River Water Resources Commission, Huazhong University of Science and Technology, China University of Geosciences, Huazhong Agricultural University, Hubei University and Jianghan University, have also carried out a lot of basic research work in the monitoring of wetland resources, ecological restoration of wetlands, research on wetland conservation system and wetland biodiversity conservation in Hubei, and and have achieved rich results, bringing lasting scientific research support power for wetland conservation in Hubei Province.

附录：湖北省湿地名录（截至2022年8月）
Appendix: List of Wetlands in Hubei Province
（as of August 2022）

【国际重要湿地名录】
List of International Important Wetlands

序号	名称	英文
1	湖北洪湖国家级自然保护区	Hubei Honghu Lake National Nature Reserve
2	湖北沉湖省级自然保护区	Hubei Chenhu Lake Provincial Nature Reserve
3	湖北神农架大九湖国家湿地公园	Hubei Shennongjia Dajiu Lake National Wetland Park
4	湖北网湖省级自然保护区	Hubei Wanghu Lake Provincial Nature Reserve

【国家重要湿地名录】
List of National Important Wetlands

序号	名称	英文
1	石首麋鹿国家重要湿地	Shishou Milu National Important Wetland
2	谷城汉江国家重要湿地	Gucheng Hanjiang River National Important Wetland
3	荆门漳河国家重要湿地	Jingmen Zhanghe National Important Wetland
4	麻城浮桥河国家重要湿地	Macheng Fuqiaohe National Important Wetland
5	潜江返湾湖国家重要湿地	Qianjiang Fanwan Lake National Important Wetland
6	松滋洈水国家重要湿地	Songzi Weishui National Important Wetland
7	武汉市安山国家重要湿地	Wuhan Anshan National Important Wetland
8	远安沮河国家重要湿地	Yuan'an Juhe River National Important Wetland

【省级重要湿地名录】
List of Provincial Important Wetlands

序号	名称	英文
1	湖北江夏上涉湖省级重要湿地	Hubei Jiangxia Shangshe Lake Provincial Important Wetland
2	湖北武汉东湖省级重要湿地	Hubei Wuhan East Lake Provincial Important Wetland
3	湖北江夏藏龙岛省级重要湿地	Hubei Jiangxia Canglong Island Provincial Important Wetland
4	湖北蔡甸后官湖省级重要湿地	Hubei Caidian Houguan Lake Provincial Important Wetland
5	湖北武汉杜公湖省级重要湿地	Hubei Wuhan Dugong Lake Provincial Important Wetland
6	湖北大冶保安湖省级重要湿地	Hubei Daye Baoan Lake Provincial Important Wetland

续表

序号	名称	英文
7	湖北襄阳汉江省级重要湿地	Hubei Xiangyang Hanjiang River Provincial Important Wetland
8	湖北樊城长寿岛省级重要湿地	Hubei Fancheng Changshou Island Provincial Important Wetland
9	湖北天门张家湖省级重要湿地	Hubei Tianmen Zhangjia Lake Provincial Important Wetland
10	湖北宜城万洋洲省级重要湿地	Hubei Yicheng Wanyangzhou Provincial Important Wetland
11	湖北老河口西排子湖省级重要湿地	Hubei Laohekou Xipaizi Lake Provincial Important Wetland
12	湖北石首天鹅洲长江故道区省级重要湿地	Hubei Shishou Swan Oxbow Yangtze River Old Road District Provincial Important Wetland
13	湖北长江新螺段白鳍豚省级重要湿地	Hubei Yangtze River Xinluo Section White-flag Dolphin Provincial Important Wetland
14	湖北公安崇湖省级重要湿地	Hubei Gong'an Chonghu Lake Provincial Important Wetland
15	湖北当阳青龙湖省级重要湿地	Hubei Dangyang Qinglong Lake Provincial Important Wetland
16	湖北枝江金湖省级重要湿地	Hubei Zhijiang Jinhu Lake Provincial Important Wetland
17	湖北长阳清江省级重要湿地	Hubei Changyang Qingjiang River Provincial Important Wetland
18	湖北竹溪万江河大鲵省级重要湿地	Hubei Zhuxi Wanjiang River Giant Salamander Provincial Important Wetland
19	湖北丹江口库区省级重要湿地	Hubei Danjiangkou Reservoir Area Provincial Important Wetland
20	湖北竹山圣水湖省级重要湿地	Hubei Zhushan Shengshui Lake Provincial Important Wetland
21	湖北竹溪龙湖省级重要湿地	Hubei Zhuxi Longhu Lake Provincial Important Wetland
22	湖北房县古南河省级重要湿地	Hubei Fangxian Gunan River Provincial Important Wetland
23	湖北十堰黄龙滩省级重要湿地	Hubei Shiyan Huanglongtan Provincial Important Wetland
24	湖北十堰郧阳湖省级重要湿地	Hubei Shiyan Yunyang Lake Provincial Important Wetland
25	湖北孝感朱湖省级重要湿地	Hubei Xiaogan Zhuhu Lake Provincial Important Wetland
26	湖北安陆府河省级重要湿地	Hubei Anlu Fuhe River Provincial Important Wetland
27	湖北云梦涢水省级重要湿地	Hubei Yunmeng Yunshui Provincial Important Wetland
28	湖北京山惠亭湖省级重要湿地	Hubei Jingshan Huiting Lake Provincial Important Wetland
29	湖北钟祥莫愁湖省级重要湿地	Hubei Zhongxiang Mochou Lake Provincial Important Wetland
30	湖北荆门仙居河省级重要湿地	Hubei Jingmen Xianju River Provincial Important Wetland
31	湖北沙洋潘集湖省级重要湿地	Hubei Shayang Panji Lake Provincial Important Wetland
32	湖北黄冈龙感湖省级重要湿地	Hubei Huanggang Longgan Lake Provincial Important Wetland
33	湖北蕲春赤龙湖省级重要湿地	Hubei Qichun Chilong Lake Provincial Important Wetland
34	湖北黄冈遗爱湖省级重要湿地	Hubei Huanggang Yiai Lake Provincial Important Wetland

续表

序号	名称	英文
35	湖北红安金沙湖省级重要湿地	Hubei Hong'an Jinsha Lake Provincial Important Wetland
36	湖北罗田天堂湖省级重要湿地	Hubei Luotian Tiantang Lake Provincial Important Wetland
37	湖北武穴武山湖省级重要湿地	Hubei Wuxue Wushan Lake Provincial Important Wetland
38	湖北浠水策湖省级重要湿地	Hubei Xishui Cehu Lake Provincial Important Wetland
39	湖北黄冈白莲河省级重要湿地	Hubei Huanggang Bailian River Provincial Important Wetland
40	湖北麻城明山省级重要湿地	Hubei Macheng Mingshan Provincial Important Wetland
41	湖北麻城三河口省级重要湿地	Hubei Macheng Sanhekou Provincial Important Wetland
42	湖北英山张家咀省级重要湿地	Hubei Yingshan Zhangjiazui Provincial Important Wetland
43	湖北赤壁陆水湖省级重要湿地	Hubei Chibi Lushui Lake Provincial Important Wetland
44	湖北通城大溪省级重要湿地	Hubei Tongcheng Daxi Provincial Important Wetland
45	湖北崇阳青山省级重要湿地	Hubei Chongyang Qingshan Provincial Important Wetland
46	湖北通山富水湖省级重要湿地	Hubei Tongshan Fushui Lake Provincial Important Wetland
47	湖北随县封江口省级重要湿地	Hubei Suixian Fengjiangkou Provincial Important Wetland
48	湖北咸丰忠建河大鲵省级重要湿地	Hubei Xianfeng Zhongjian River Giant Salamander Provincial Important Wetland
49	湖北咸丰二仙岩省级重要湿地	Hubei Xianfeng Erxianyan Provincial Important Wetland
50	湖北宣恩贡水河省级重要湿地	Hubei Xuanen Gongshui River Provincial Important Wetland
51	湖北仙桃沙湖省级重要湿地	Hubei Xiantao Shahu Lake Provincial Important Wetland
52	湖北梁子湖省级重要湿地	Hubei Liangzi Lake Provincial Important Wetland

【国家级自然保护区名录】
List of National Nature Reserves

序号	名称	英文
1	湖北龙感湖国家级湿地自然保护区	Hubei Longgan Lake National Wetland Nature Reserve
2	湖北洪湖国家级自然保护区	Hubei Honghu Lake National Nature Reserve
3	湖北长江新螺段白鱀豚国家级自然保护区	Hubei Yangtze River Xinluo Section White-flag Dolphin National Wetland Nature Reserve
4	湖北长江天鹅洲白鱀豚国家级自然保护区	Hubei Yangtze River Swan Oxbow White-flag Dolphin National Nature Reserve
5	湖北石首麋鹿国家级自然保护区	Hubei Shishou Milu National Nature Reserve
6	湖北咸丰忠建河大鲵国家级自然保护区	Hubei Xianfeng Zhongjian River Giant Salamander National Nature Reserve

【省级自然保护区名录】
List of Provincial Nature Reserves

序号	名称	英文
1	湖北网湖湿地自然保护区	Hubei Wanghu Lake Wetland Nature Reserve
2	湖北武汉沉湖湿地省级自然保护区	Hubei Wuhan Chenhu Wetland Provincial Nature Reserve
3	湖北梁子湖省级湿地自然保护区	Hubei Liangzi Lake Provincial Wetland Natural Reserve
4	湖北丹江口库区省级湿地自然保护区	Hubei Danjiangkou Reservoir Provincial Wetland Natural Reserve
5	神农架大九湖湿地省级自然保护区	Shennongjia Dajiu Lake Wetland Provincial Natural Reserve
6	咸丰二仙岩湿地省级自然保护区	Xianfeng Erxianyan Wetland Provincial Nature Reserve
7	上涉湖湿地省级自然保护区	Shangshe Lake Wetland Provincial Nature Reserve
8	湖北万江河大鲵自然保护区	Hubei Wanjiang River Giant Salamander Nature Reserve
9	湖北长江宜昌中华鲟省级自然保护区	Hubei Yangtze River Yichang Chinese Sturgeon Provincial Nature Reserve
10	何王庙长江江豚省级自然保护区	Hewangmiao Yangtze Finless Porpoise Provincial Nature Reserve

【国家湿地公园名录】
List of National Wetland Parks

序号	名称	英文
1	湖北武汉东湖国家湿地公园	Hubei Wuhan East Lake National Wetland Park
2	湖北江夏藏龙岛国家湿地公园	Hubei Jiangxia Canglong Island National Wetland Park
3	湖北武汉安山国家湿地公园	Hubei Wuhan Anshan National Wetland Park
4	湖北蔡甸后官湖国家湿地公园	Hubei Caidian Houguan Lake National Wetland Park
5	湖北武汉杜公湖国家湿地公园	Hubei Wuhan Dugong Lake National Wetland Park
6	湖北谷城汉江国家湿地公园	Hubei Gucheng Hanjiang River National Wetland Park
7	湖北樊城长寿岛国家湿地公园	Hubei Changshou Island National Wetland Park
8	湖北襄阳汉江国家湿地公园	Hubei Xiangyang Hanjiang River National Wetland Park
9	湖北宜城万洋洲国家湿地公园	Hubei Yicheng Wanyangzhou National Wetland Park
10	湖北南漳清凉河国家湿地公园	Hubei Nanzhang Qingliang River National Wetland Park
11	湖北老河口西排子湖国家湿地公园	Hubei Laohekou Xipaizi Lake National Wetland Park
12	湖北宜都天龙湾国家湿地公园	Hubei Yidu Tianlongwan National Wetland Park
13	湖北当阳青龙湖国家湿地公园	Hubei Dangyang Qinglong Lake National Wetland Park
14	湖北远安沮河国家湿地公园	Hubei Yuan'an Juhe River National Wetland Park
15	湖北长阳清江国家湿地公园	Hubei Changyang Qingjiang River National Wetland Park

续表

序号	名称	英文
16	湖北枝江金湖国家湿地公园	Hubei Zhijiang Jinhu Lake National Wetland Park
17	湖北夷陵圈椅淌国家湿地公园	Hubei Yiling Quanyitang National Wetland Park
18	湖北五峰百溪河国家湿地公园	Hubei Wufeng Baixi River National Wetland Park
19	湖北秭归九畹溪国家湿地公园	Hubei Zigui Jiuwanxi National Wetland Park
20	湖北大冶保安湖国家湿地公园	Hubei Daye Baoan Lake National Wetland Park
21	湖北阳新莲花湖国家湿地公园	Hubei Yangxin Lianhua Lake National Wetland Park
22	湖北竹山圣水湖国家湿地公园	Hubei Zhushan Shengshui Lake National Wetland Park
23	湖北竹溪龙湖国家湿地公园	Hubei Zhuxi Longhu Lake National Wetland Park
24	湖北房县古南河国家湿地公园	Hubei Fangxian Gunan River National Wetland Park
25	湖北十堰黄龙滩国家湿地公园	Hubei Shiyan Huanglongtan National Wetland Park
26	湖北十堰郧阳湖国家湿地公园	Hubei Shiyan Yunyang Lake National Wetland Park
27	湖北十堰泗河国家湿地公园	Hubei Shiyan Sihe River National Wetland Park
28	湖北环荆州古城国家湿地公园	Hubei Huanjingzhou Ancient City National Wetland Park
29	湖北松滋洈水国家湿地公园	Hubei Songzi Weishui National Wetland Park
30	湖北荆州菱角湖国家湿地公园	Hubei Jingzhou Lingjiao Lake National Wetland Park
31	湖北石首三菱湖国家湿地公园	Hubei Shishou Sanling Lake National Wetland Park
32	湖北监利老江河故道国家湿地公园	Hubei Jianli Laojiang River Old Road National Wetland Park
33	湖北公安崇湖国家湿地公园	Hubei Gong'an Chonghu Lake National Wetland Park
34	湖北荆门漳河国家湿地公园	Hubei Jingmen Zhanghe River National Wetland Park
35	湖北荆门仙居河国家湿地公园	Hubei Jingmen Xianjuhe National Wetland Park
36	湖北京山惠亭湖国家湿地公园	Hubei Jingshan Huiting Lake National Wetland Park
37	湖北钟祥莫愁湖国家湿地公园	Hubei Zhongxiang Mochou Lake National Wetland Park
38	湖北沙洋潘集湖国家湿地公园	Hubei Shayang Panji Lake National Wetland Park
39	湖北安陆府河国家湿地公园	Hubei Anlu Fuhe River National Wetland Park
40	湖北孝感朱湖国家湿地公园	Hubei Xiaogan Zhuhu Lake National Wetland Park
41	湖北孝感老观湖国家湿地公园	Hubei Xiaogan Laoguan Lake National Wetland Park
42	湖北汉川汈汊湖国家湿地公园	Hubei Hanchuan Diaocha Lake National Wetland Park
43	湖北云梦涢水国家湿地公园	Hubei Yunmeng Yunshui National Wetland Park
44	湖北蕲春赤龙湖国家湿地公园	Hubei Qichun Chilong Lake National Wetland Park
45	湖北黄冈遗爱湖国家湿地公园	Hubei Huanggang Yiai Lake National Wetland Park
46	湖北红安金沙湖国家湿地公园	Hubei Hong'an Jinsha Lake National Wetland Park

续表

序号	名称	英文
47	湖北罗田天堂湖国家湿地公园	Hubei Luotian Tiantang Lake National Wetland Park
48	湖北武穴武山湖国家湿地公园	Hubei Wuxue Wushan Lake National Wetland Park
49	湖北麻城浮桥河国家湿地公园	Hubei Macheng Fuqiao River National Wetland Park
50	湖北浠水策湖国家湿地公园	Hubei Xishui Cehu Lake National Wetland Park
51	湖北黄冈白莲河国家湿地公园	Hubei Huanggang Bailian River National Wetland Park
52	湖北英山张家咀国家湿地公园	Hubei Yingshan Zhangjiazui National Wetland Park
53	湖北通城大溪国家湿地公园	Hubei Tongcheng Daxi National Wetland Park
54	湖北嘉鱼珍湖国家湿地公园	Hubei Jiayu Zhenhu Lake National Wetland Park
55	湖北咸宁向阳湖国家湿地公园	Hubei Xianning Xiangyang Lake National Wetland Park
56	湖北崇阳青山国家湿地公园	Hubei Congyang Qingshan National Wetland Park
57	湖北赤壁陆水湖国家湿地公园	Hubei Chibi Lushui Lake National Wetland Park
58	湖北通山富水湖国家湿地公园	Hubei Tongshan Fushui Lake National Wetland Park
59	湖北随县封江口国家湿地公园	Hubei Suixian Fengjiangkou National Wetland Park
60	湖北广水徐家河国家湿地公园	Hubei Guangshui Xujia River National Wetland Park
61	湖北随州淮河国家湿地公园	Hubei Suizhou Huaihe River National Wetland Park
62	湖北宣恩贡水河国家湿地公园	Hubei Xuanen Gongshui River National Wetland Park
63	湖北仙桃沙湖国家湿地公园	Hubei Xiantao Shahu Lake National Wetland Park
64	湖北天门张家湖国家湿地公园	Hubei Tianmen Zhangjia Lake National Wetland Park
65	湖北潜江返湾湖国家湿地公园	Hubei Qianjiang Fanwan Lake National Wetland Park
66	湖北神农架大九湖国家湿地公园	Hubei Shennongjia Dajiu Lake National Wetland Park

【省级湿地公园名录】
List of Provincial Wetland Parks

序号	名称	英文
1	湖北蔡甸索子长河省级湿地公园	Hubei Caidian Suozichang River Provincial Wetland Park
2	湖北蔡甸桐湖省级湿地公园	Hubei Caidian Tonghu Lake Provincial Wetland Park
3	湖北江夏潴洋海省级湿地公园	Hubei Jiangxia Zhuyanghai Provincial Wetland Park
4	湖北黄陂木兰花溪省级湿地公园	Hubei Huangpi Mulan Huaxi Provincial Wetland Park
5	湖北崔家营省级湿地公园	Hubei Cuijiaying Provincial Wetland Park
6	湖北枣阳熊河省级湿地公园	Hubei Zaoyang Xionghe River Provincial Wetland Park
7	湖北宜城鲤鱼湖省级湿地公园	Hubei Yicheng Liyu Lake Provincial Wetland Park

续表

序号	名称	英文
8	湖北枝江玛瑙河省级湿地公园	Hubei Zhijiang Manao River Provincial Wetland Park
9	湖北江陵龙渊湖省级湿地公园	Hubei Jiangling Longyuan Lake Provincial Wetland Park
10	湖北监利锦沙湖省级湿地公园	Hubei Jianli Jinsha Lake Provincial Wetland Park
11	湖北石首山底湖省级湿地公园	Hubei Shishou Shandi Lake Provincial Wetland Park
12	湖北公安淤泥湖省级湿地公园	Hubei Gong'an Yuni Lake Provincial Wetland Park
13	湖北洪湖新滩省级湿地公园	Hubei Honghu Lake Xintan Provincial Wetland Park
14	湖北京山石龙水库省级湿地公园	Hubei Jingshan Shilong Reservoir Provincial Wetland Park
15	湖北钟祥石门湖省级湿地公园	Hubei Zhongxiang Shimen Lake Provincial Wetland Park
16	湖北荆门钱河省级湿地公园	Hubei Jingmen Qianhe River Provincial Wetland Park
17	湖北荆门象河省级湿地公园	Hubei Jingmen Xianghe River Provincial Wetland Park
18	湖北掇刀官冲省级湿地公园	Hubei Duodao Guanchong Provincial Wetland Park
19	湖北屈家岭青木垱河省级湿地公园	Hubei Qujialing Qingmudang River Provincial Wetland Park
20	湖北鄂州市洋澜湖省级湿地公园	Hubei Ezhou Yanglan Lake Provincial Wetland Park
21	湖北大悟九房沟省级湿地公园	Hubei Dawu Jiufanggou Provincial Wetland Park
22	湖北孝昌观音湖省级湿地公园	Hubei Xiaochang Guanyin Lake Provincial Wetland Park
23	湖北黄州道仁湖省级湿地公园	Hubei Huangzhou Daoren Lake Provincial Wetland Park
24	湖北黄州滨江省级湿地公园	Hubei Huangzhou Riverside Provincial Wetland Park
25	湖北红安倒水河省级湿地公园	Hubei Hong'an Daoshui River Provincial Wetland Park
26	湖北麻城明山省级湿地公园	Hubei Macheng Mingshan Provincial Wetland Park
27	湖北麻城三河口省级湿地公园	Hubei Macheng Sanhekou Provincial Wetland Park
28	湖北罗田跨马墩省级湿地公园	Hubei Luotian Kuamadun Provincial Wetland Park
29	湖北罗田义水河省级湿地公园	Hubei Luotian Yishui River Provincial Wetland Park
30	湖北蕲春仙人湖省级湿地公园	Hubei Qichun Xianren Lake Provincial Wetland Park
31	湖北武穴长江外滩省级湿地公园	Hubei Wuxue Yangtze River Bund Provincial Wetland Park
32	湖北武穴梅川省级湿地公园	Hubei Wuxue Meichuan Provincial Wetland Park
33	湖北武穴荆竹省级湿地公园	Hubei Wuxue Jingzhu Provincial Wetland Park
34	湖北武穴大金省级湿地公园	Hubei Wuxue Dajin Provincial Wetland Park
35	湖北武穴仙人坝省级湿地公园	Hubei Wuxue Xianren Dam Provincial Wetland Park
36	湖北咸安金桂湖省级湿地公园	Hubei Xian'an Jingui Lake Provincial Wetland Park
37	湖北通山望江岭省级湿地公园	Hubei Tongshan Wangjiangling Provincial Wetland Park
38	湖北崇阳浪口省级湿地公园	Hubei Chongyang Langkou Provincial Wetland Park